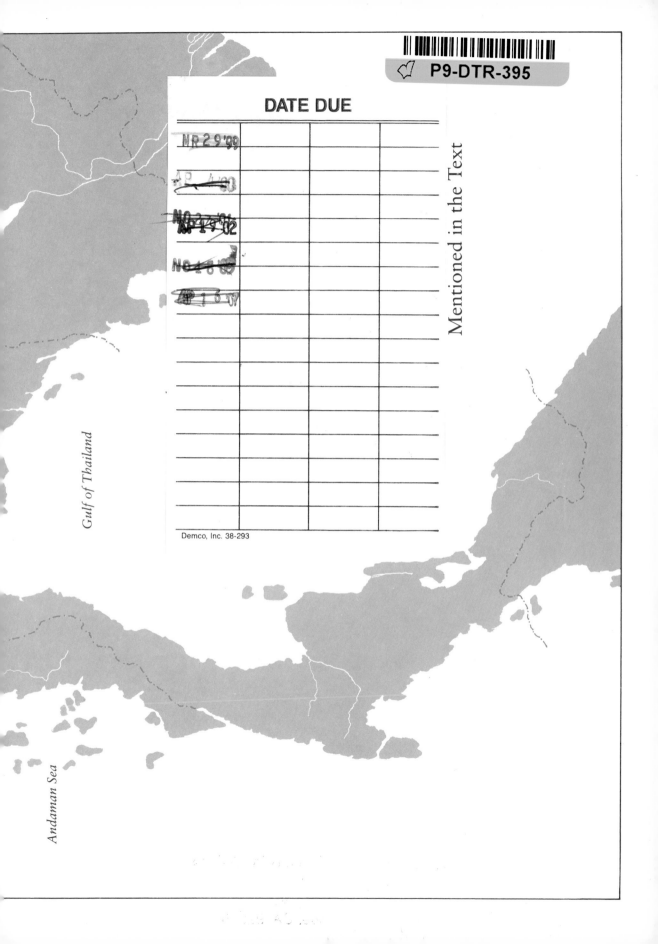

DATE DUE

MR 29 '99			
AP 4 '00			
NO 27 '01 AP 17 '02			
NO 18 '02			
AP 1 8 '07			

Demco, Inc. 38-293

Mentioned in the Text

Gulf of Thailand

Andaman Sea

ELEPHANTS OF THAILAND
IN MYTH, ART, AND REALITY

Elephants of Thailand in Myth, Art, and Reality

Rita Ringis

Kuala Lumpur
OXFORD UNIVERSITY PRESS
Oxford   Singapore   New York
1996

Oxford University Press

Oxford New York
Athens Auckland Bangkok Bombay
Calcutta Cape Town Dar es Salaam Delhi
Florence Hong Kong Istanbul Karachi
Madras Madrid Melbourne Mexico City
Nairobi Paris Shah Alam Singapore
Taipei Tokyo Toronto

and associated companies in
Berlin Ibadan

Oxford is a trade mark of Oxford University Press

Published in the United States
by Oxford University Press, New York

British Library Cataloguing in Publication Data
Data available

Library of Congress Cataloging-in-Publication Data
Ringis, Rita, 1942–
    Elephants of Thailand in myth, art, and reality/Rita Ringis.
       p.   cm.
    Includes bibliographical references and index.
    ISBN 967 65 3068 9
1. Elephants—Thailand. 2. Elephants—Thailand—Mythology.
3. Elephants in art.   I. Title.
QL737 P98R56 1996
599.6'1'09593—dc20
95–40821
CIP

Typeset by EXPO Holdings Sdn. Bhd., Malaysia
Printed by KHL Printing Co. (S) Pte. Ltd., Singapore
Published by the South-East Asian Publishing Unit,
a division of Penerbit Fajar Bakti Sdn. Bhd.,
under licence from Oxford University Press,
4 Jalan U1/15, Seksyen U1, 40000 Shah Alam,
Selangor Darul Ehsan, Malaysia

Preface and Acknowledgements

WRITING about elephants would seem to be an odd if not presumptuous undertaking for someone whose major field of interest is in Thai art history and culture. As a resident of Thailand for some fourteen years until 1991, I was most fortunate to have had the opportunity to experience at first hand many aspects of that culture as well as to study it in some considerable depth, and in due course, over the years, to convey to others, through guiding, lecturing, and writing, some of the insights I had gained. Thus my temerity in venturing to eulogize the Thai elephant was prompted by a desire to refresh the memories of those who may have taken its role for granted, and to introduce those unaware of it, to the literally immense presence of the elephant in Thai life of the past.

Throughout Thai history, the nation has drawn for its strength, both literally and metaphorically, on its seemingly endless reserves of elephants, once truly synonymous with the land and its people. Perhaps because this notion is still so deeply ingrained in the mind-set of the Thai people, so much an accepted feature, be it in the language, the arts, or history, it is no longer considered remarkable. However, what is remarkable is the fact that at present the elephant, this resource of national tradition and pride, is in grave danger of extinction, of becoming in the future no more than one of the many literally fabulous mythical creatures whose effigies ornament Thai temples as guardians of Buddhism.

While touching upon the grim statistics of this serious problem, I leave their further enumeration and clarification in the capable hands of the specialist scientists and researchers currently working in the field. My aim is to take the reader on a journey, on elephant-back as it were, through some of the highways and byways of Thai history, art, and culture. To that end, each chapter in this book is intended to be read as a single unit, a window on Thailand, past and present.

The compilation of this book, rather akin to the lengthy gestation period of an elephant, has taken some considerable time. Since undertaking its writing, I have lived and travelled intermittently between three different countries—Thailand, Singapore, and Australia—and in each of these, numerous institutions, officials, and friends have generously responded to my requests for assistance.

Thus I have many to thank for their generous help in my seemingly endless elephantine enterprise. However, I must emphasize that any errors of interpretation of the varied information kindly provided by them are my own.

In Thailand, my debts of gratitude are considerable. First and foremost, I feel greatly privileged to have had the opportunity to benefit, over the years, from the teachings of His Serene Highness Prince Subhadradis Diskul, Rector Emeritus of Silpakorn Fine Arts University, who has been a constant source of inspiration and encouragement to generations of students and *amateurs* of the arts of Thailand. In my own undertakings in writing about Thailand, I am deeply indebted to him for his kind guidance and counsel.

In this more recent venture, I am greatly beholden to Mom Luang Thawisan Ladawan, His Majesty's Principal Private Secretary, for graciously arranging Royal Permission for me to visit the Royal Elephant Stables at Chitralada Palace, and to Mom Rachawong Putrie Viravaidya of the Bureau of the Royal Household for her invaluable assistance. My wholehearted thanks go also to Dr Mom Luang Phiphatanachatr Diskul, and to Khun Svest Dhanapradith.

I am also most grateful to Dr Suvit Rasmibhuti, Director-General of the Fine Arts Department, and to Khun Thiti Burakamkovit, Deputy Director-General, for permission to use illustrative materials. At the National Library of Thailand, I thank the Director, Khun Thara Kanakamani, and Chief of the Manuscripts Section, Khun Niyadah Tasukon. My thanks go also to the Director and Staff of the National Museum in Bangkok.

At the Forest Industry Organization, I thank the following for assistance, information, and illustrative material: the Managing Director, Colonel Mom Rachawong Aduldej Chakrabandhu, and particularly Khun Amnuay Sorachart and Khun Mookda Nititoolteppagul. Greatly appreciated as well was the informative help given by Dr Pricha Puangkham, Head of the Young Elephant Training School at Ngao, outside of Lampang, where Richard Lair was also most helpful. In addition, I am most grateful to Khun Patamavadee Ratacharern in Lampang.

My thanks go also to the Director of the Wildlife Conservation Division of the Royal Thai Forest Department, and to Khun Mattana Srikrachang for her assistance. The Wildlife Fund Thailand under the Royal Patronage of Her Majesty the Queen was also most helpful with information and material, and I thank its Secretary-General, and Khun Suraphon Duangkhae, Projects Director. Illustrative material for this publication was also kindly provided by the Tourism Authority of Thailand, Bangkok Head Office.

In addition, I am more than indebted to a remarkable institution in Bangkok, the National Museum Volunteers Group, for stimulation and opportunities afforded me over the years, and for providing the foundations for my present endeavour. My most grateful thanks go to many National Museum Volunteer colleagues, both past and present, for their help and companionship in the quest for knowledge and understanding. On a more personal level in this present undertaking, for generous hospitality and continuing encouragement both in word and deed, I am greatly indebted to Eileen and Julian Deeley, Monique and Hermann Heitmann, Ruth and Michael Gerson, and Fumiko and Robert Boughey. I am also most grateful to Acharn Smitthi Siribhadra.

My thanks are also extended to the Siam Society and its President, as well as to the editor of its journal, James V. Di Crocco, and to Khun Euayporn Kerdchuay. Additionally, I am most grateful to Dacre F. A. Raikes, OBE, former Vice-President of the Society.

I am greatly obliged to Nicky Van Oudenhoven, Wanna Sasitanond, Sarah McLean, Ann White, Marjorie Manley, Virginia Sriporamanont, Roger Welty of Eastern Horizons, and Karen Chungyampin, Curator of the Tilleke and Gibbins Collection of South-East Asian Textiles.

In Singapore, my most appreciative thanks go to Carol and Win Mumby, whose wonderful generosity and friendship greatly sustained my morale during the actual writing of this book. Heartfelt thanks are also due to Ginny and Clifford Granger, Helene and David Haines, Andu Dixon, and Sirajuddin Ismail. For their good companionship during that time, I also thank the Thursday Group. As well, the following organizations were most helpful: the staff of the South-East Asian Section of the Singapore National Library, and the staff at the library of ISEAS (Institute of South-East Asian Studies). At the Singapore Zoological Gardens, I am most appreciative of the help given me by Kumar Pillai, particularly in arranging for me to meet the Chairman of the IUCN/SSC Asian Elephant Specialist Group, Lyn de Alwis, whose words of encouragement and information were most welcome.

In Sydney, my thanks go to the following for their helpful encouragement: Jennifer Isaacs, Peter and Inese King, and Peter Ross. The officers of the Sydney Branch of the Tourism Authority of Thailand were of great assistance, and there I thank Khun Saichalee Varnapruk, Khun Chattan Kunjara na Ayudhya, Isabel Ringis, and Neil Ainsworth. I also thank Lek McFadden of the National Library in Canberra, and in Perth, I am most grateful to Carolyn Dupont, Ken Whitbread, and Nongyao Premkamolnetr.

In addition I would like to thank the following National Museum Volunteer colleagues no longer living in Thailand for their continuing generous assistance with illustrative material: Kim Retka of Mexico City, Pam Taylor of the United States, Barbara Rowbottom of Holland, Libby Kugler of Indonesia, Elizabeth Dhe of France, and Paulette de Schaller of Switzerland. Thanks go also to Karen Kane of the Metro Washington Park Zoo, Portland, Oregon, for providing informative material. My warmest thanks are also extended to the editors at Oxford University Press in Kuala Lumpur for their patience and encouragement.

And finally, for their equanimity, patience, and forbearance, virtues sorely tested during the writing of this book, I am, as always, indebted to my husband John and my children, Isabel and Alexander.

Perth, Western Australia   RITA RINGIS
October 1995

Contents

Plates

Figures

Note on Terminology

THAILAND has been the official international name of the country since 1939. This is in fact the English translation of the term by which the Thai people designate their land and race, *Muang Thai* or *Prathet Thai*, the Land of the Thai.

Prior to 1939 the country was popularly known as Siam, and the people as Siamese. While the words 'Thai' and 'Siam' are of ancient origin, from the sixteenth century onwards, 'Siam' and 'Siamese' were the terms used by Europeans when writing of the land and its people.

In this book, the terms 'Siam' and 'Siamese' are used when referring to historical events and characters between that time and the mid-twentieth century. However, the terms 'Thailand' and 'Thai' are used throughout (irrespective of time frame) to designate the geographic areas encompassed within current national boundaries, and the people predominant there since the late thirteenth century.

1 Introduction

ARCHAEOLOGICAL findings indicate that since earliest recorded time, people have been fascinated by the size, strength, and sheer grandeur of the elephant, the largest living land mammal. For the current two surviving species of the elephant family, the African (*Loxodonta africana*) and the Asian (*Elephas maximus*), neither size, nor strength, nor least of all our fickle admiration, will be of much relevance in the future, as the very existence of both species is currently threatened. Present-day African elephants number fewer than one million. Asian elephants, classed as endangered, are distributed in scattered at-risk populations in Bangladesh, Bhutan, Burma, Cambodia, China, India, Indonesia (Kalimantan and Sumatra), Laos, Malaysia (Peninsular Malaysia and Sabah), Nepal, Sri Lanka, Thailand, and Vietnam. Their total numbers are estimated at a bare one-tenth of the African elephant populations, being optimistically assessed as ranging between 30,000 and 55,000 (Santiapillai and Jackson, 1990: v). Both species, though evolved in vastly different conditions and climates, face similar and closely linked problems.

In Africa and Asia alike, the perennial requirements of land safe for varied husbandry and development, multiplied by present-day human population pressures, of necessity continue to confine and press dwindling elephant populations towards extinction. This process is compounded by the continuing world demand for ivory trinkets. Although trade in ivory is restricted in countries that are signatories of the Convention on International Trade in Endangered Species (CITES), effective enforcement is difficult. Thus while legislation against poaching and untrammelled trade in elephant products is being variously implemented in Africa, both appear to continue on a reduced and illegal scale. To supplement diminished supplies of African ivory for the world market, it seems that Asian elephants are currently being hunted for ivory, thus depleting the gene pool of tusked elephants. Unlike the African elephants, not all Asian elephants carry tusks, these being confined to some males, but not all. Further contributing to the decline of the Asian elephant species is the burgeoning trade in South-East Asia in elephant hides for leather goods manufacture.

Of great significance to those concerned in perpetuating the survival of the Asian elephant are the detailed comparative findings and recommendations for the future, documented in

1

Santiapillai and Jackson's *The Asian Elephant: An Action Plan for Its Conservation,* a report produced in 1990 by the IUCN/SSC Asian Elephant Specialist Group and published by the International Union for Conservation of Nature and Natural Resources. Financially assisted by various international bodies and zoological societies, the IUCN/SSC Asian Elephant Specialist Group (AESG) provides a regional forum for the promotion of economically and ecologically realistic elephant conservation in Asia.[1] The AESG Action Plan includes recommendations for environmentally responsible and economically feasible development planning. This takes into account pragmatic conservation and management of elephant populations within the larger context of other natural resources utilization. Also stressed is the need for protective legislation and its effective enforcement for conservation.

While recognizing that such measures can only be initiated from within the political arenas of countries with elephant populations, the report notes that the elephant is a part of all mankind's heritage, and thus urges co-operation from the rest of the world, whether in spirit through strict adherence to international agreements that promote responsibility towards elephant products (CITES), or with actual financial assistance aimed at conservation programmes initiated within countries with elephant populations (Santiapillai and Jackson, 1990: 1–8).

[1]The Group includes representatives from Bangladesh, Bhutan, Burma, Cambodia, China, India, Indonesia, Laos, Malaysia, Nepal, Sri Lanka, Vietnam, and Thailand.

Fig. 1
Mahout and tame elephants near Petchaburi. Engraving from Henri Mouhot, *Travels in the Central Parts of Indo-China...*, Vol. 1, 1864.

Until the recent past, as home to vast numbers of the Asian elephant, Thailand, formerly known as the kingdom of Siam, was unrivalled. Traditional and seemingly extravagant Siamese (Thai) claims as to the richness of their elephant heritage were in fact largely confirmed in observations made by foreign travellers and foreign merchants resident in the kingdom, particularly between the seventeenth and the early twentieth centuries. These recorded that extensive herds of wild elephants roamed free in lush forest jungles, or captive and trained, in their thousands worked beside man in the fields of commerce, transport, and war (Plates 1 and 2; Fig. 1). Particularly notable also was the role played by elephants in the kingdom's state processions and pageantry (Fig. 2).

Doubtless over the centuries, the continuing supply of these thousands of working elephants derived not only from those born and bred in captivity, but also from the training of wild elephants captured during traditional large-scale annual elephant drives held in various parts of the country. Although these hunts had effectively ceased by the early twentieth century, the forests that still covered much of the land continued to harbour an abundance of wild herds. As a result, elephants were frequent predators on cultivated fields which then lay close to the forest jungles. The havoc perennially wreaked by rampaging elephants on villagers of former times was vividly captured for posterity on film during the 1930s, in a black and white silent movie called 'Chang!' ('Elephant!'), preserved at the National Film Archives of Thailand.

Fig. 2
Procession with elephants and retinue which includes foreign mercenaries. Detail from a modern facsimile replica of a *khoi* paper manuscript copy of seventeenth-century murals at Wat Yom, Ayutthaya. (Photograph courtesy of the Fine Arts Department, Bangkok)

In the present day, such attacks from wild elephants are still a painful reality in some Asian countries with elephant populations. However, in Thailand the elephant is rarely a predator today, being confined in decreasing numbers to relatively isolated rugged forest areas.

In contrast to the seemingly inexhaustible profusion of elephants in the past, a recent optimistic 'guesstimation', given in 1994 at a seminar on the future of the Thai elephant, puts their total numbers at some 3,000, of which about half are domestic elephants, living in captivity (Sapana Sakya, 1994). However, this decline in elephant populations in Thailand, unlike that in Africa, has had for the most part little to do with large-scale trade in ivory. For centuries prior to the twentieth, the Thai elephant's safety was ensured by long-standing taboos invoking religious and traditional beliefs. The elephant was held to be divinely created, and thus associated with the works of the gods (see Fig. 35). More importantly, its white variation (Fig. 3) was revered as a sacred harbinger in the life of the Buddha (see Plate 25). Thus the elephant was not hunted for meat, and was never regarded as a mere producer of material that could be fashioned into luxury objects and trinkets for profit. Few major works of art (until recently by definition religious) in Thailand were fashioned from ivory.

In Thai, the term for 'ivory' is *nga chang*, meaning 'the tusk of the elephant', a term that directly relates to a living, breathing animal, unlike the word 'ivory' which distances itself from the creature that produces it, allowing the possessor of ivory objects the luxury of a clean if negligent conscience. Indeed, the traditional Thai odium at slaughter for ivory is clearly inherent in a proverb that warns against sacrificing a lot to gain a little: 'Kah chang ao nga' (literally, kill an elephant for its tusks). Thus it would be reasonable to assume that the relatively small amounts of ivory exported in past centuries came from the customary prac-

Fig. 3
A white elephant. Nineteenth-century *khoi* paper manuscript. (Photograph courtesy of the National Library, Bangkok)

4

tice of tipping or cutting off the tusks of the captive elephant without resorting to killing the animal (see Fig. 41). After all, even in purely commercial terms, the elephant was of greater value as a beast of burden or vehicle of war. Over and above traditional sanctions, legislation introduced in the early twentieth century prohibited unauthorized hunting, killing, or even wounding of elephants in the wild.

Paradoxically, Thailand's potential loss of its ancient elephant heritage has been closely associated with economic development, in the extensive exploitation of its forests during the twentieth century. The Thai (Asian) elephant, unlike its sun-loving African cousin, can survive and proliferate only in shady forest jungle areas, protected from the hot climate, and sustained by that same forest vegetation (Fig. 4). Until well into the twentieth century, Thailand was an agriculturally dependent country with a rapidly increasing population. Thus forested areas of necessity gave way to increasing land cultivation. Concurrent with this was the rapid expansion of the timber trade, of great importance to the country's economic development throughout the greater part of the twentieth century. In this wealth-producing industry, trained working elephants were integral partners, labouring beside man to extract the green gold from the forests, the habitat of the wild elephant.

From the late 1960s onwards, in order to redress this environmental depletion, various conservation and reforestation programmes have been progressively introduced by successive governments. By 1990 forest areas designated and protected as national parks and wildlife sanctuaries covered some 10 per cent of Thailand's land area, with 15 per cent conservation forest nation-wide as the aim for the future (Wildlife Fund Thailand, 1990: 5). It is largely in these protected areas that the Thai elephant survives (Fig. 5).

While comprehensive country-wide assessments of the numbers of elephants in the wild in Thailand have as yet to be fully implemented, studies undertaken in known elephant areas since the 1970s confirmed a pattern of steady decline in wild elephant numbers concurrent with the corresponding decline of forested land. Cited in the AESG report mentioned earlier, Boonsong Legakul and McNeely's study (1977) of the major elephant supporting areas of Thailand (including wildlife sanctuaries and parks) estimated wild elephant numbers as ranging from a minimum of 2,600 to a maximum of 4,450. Drawing attention to the more recent figures of Dobias (1987), the report noted a further decline, with a maximum of only 1,700 elephants in protected Wildlife Sanctuaries and National Parks, and with some elephants estimated to be outside those protected areas (Santiapillai and Jackson, 1990: 67). While any reasonable lay person would regret such a decline, few would be aware of the genetic consequences of these diminished populations of elephants in Thailand. However, the AESG report noted that to maintain

Fig. 4
Set of four postage stamps from the Post Office of Thailand depicting (i) elephants working in forest, (ii) elephant kneeling, (iii) elephants bathing, and (iv) elephant mother and calves.

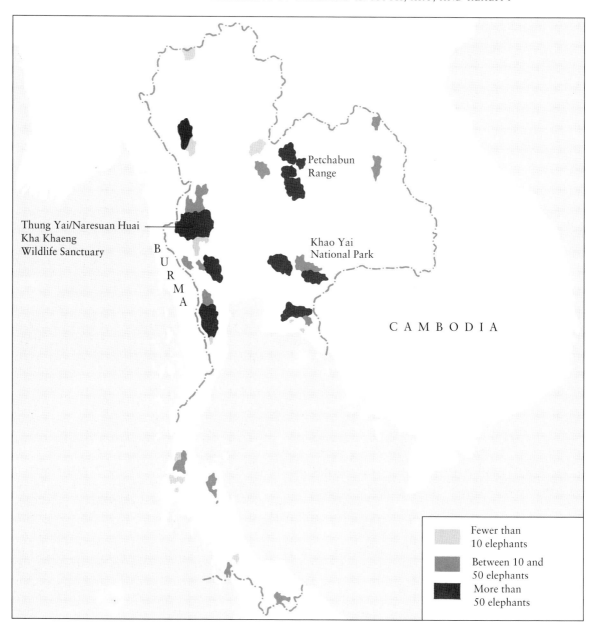

Thung Yai/Naresuan Huai
Kha Khaeng
Wildlife Sanctuary

Petchabun
Range

Khao Yai
National Park

B
U
R
M
A

C A M B O D I A

Fewer than
10 elephants

Between 10 and
50 elephants

More than
50 elephants

Fig. 5
Map of in-progress assessment of wild
elephant population distributions in
Thailand, 1990. (Courtesy of Mattana
Srikrachang, Wildlife Conservation
Division of Royal Thai Forest
Department)

the potential for continued evolution based on natural selection, a
single population of at least 2,000 elephants was necessary.
Acknowledging that habitat fragmentation had rendered that
optimum figure unrealistic, the report recommended management
plans suitable for maintaining the viability of smaller populations
(Santiapillai and Jackson, 1990: 4–5; 67–72).

The report also noted that in the various protected areas in
Thailand, the largest populations (comprising many small herds)
were estimated to total some 250 elephants in the Khao Yai
National Park, an area covering 2168 square kilometres and
extending into four provinces, and some 300 elephants in the

6

heavily forested wildlife sanctuaries at Huai Kha Khaeng and Thung Yai Naresuan covering some 5000 square kilometres in the Kanchanaburi and Tak provinces. This latter area was assessed as 'perhaps the only area in Thailand capable of sustaining a viable elephant population in the long term' under then current conditions of land use (Santiapillai and Jackson, 1990: 72). In fact in Thailand, 'present day [wild] elephant populations are, for the most part, small, isolated and declining rapidly in an environment dominated by man' (Santiapillai and Jackson, 1990: 67).

Here is the crux of the problem, man and his needs as against those of the creatures of the forest and the forest itself. These issues were vividly illustrated in a documentary film series, screened nation-wide in Thailand to stimulate public awareness of the gravity of the problems caused by continuing environmental degradation. Produced by the Wildlife Fund Thailand[2] in co-operation with the Siam Commercial Bank, the series was entitled 'Elephant, the Forgotten Friend'. While documenting the Thai elephant's undoubted place in the cultural life of the past, the series highlighted the elephant's pivotal role in the conservation and regeneration of forest habitats in the present.

Conservationists recognize the elephant as an 'indicator species', one of the first to suffer as a result of habitat fragmentation (Mattana Srikrachang et al., 1991). Decline of elephant population is in fact a warning, an indication of corresponding decline in other species that share its habitat and benefit directly from its presence, for within the forest the elephant, man's 'Forgotten Friend', plays a key role in maintaining the delicate balance between the many interdependent species of plants and animals.

It is common knowledge, for example, that elephants consume vast amounts of fodder daily, approximately 200 kilograms per head. Thus the consumption pattern of a herd of some twenty elephants in any given area of feeding would of necessity be considerable. Contrary to the immediate assumption that such feeding patterns must be destructive, within a healthy and viable habitat they are actually constructive and promote regeneration. When the elephant feeds on high branches in the forest, it provides openings for sunlight to penetrate to low growing plants. Being a discriminating eater, it discards various roots or branches which on decay provide rich humus for the soil, or, if fresh, nourishment for smaller forest creatures. These creatures also benefit from the 'tunnels' made in the undergrowth by the elephant's passing. In other regular activities such as digging for roots, snuffling at natural salt-licks, and even sharpening its tusks, the elephant turns over the soil, liberating useful bacteria. Even its dung has value in the forest, for the elephant's large-scale on-the-move consumption of fodder and its subsequent elimination scatters plant seeds for

[2]Founded in 1983 as a non-profit organization, the Wildlife Fund Thailand under the Royal Patronage of Her Majesty the Queen sponsors many pragmatic projects aimed at protection and conservation of the natural environment.

regrowth. In fact, as pointed out in the AESG Action Plan, any conservation programme aimed specifically at the elephant would 'ensure the maintenance of biological diversity and ecological integrity on a large scale' (Santiapillai and Jackson, 1990: vii).

However, despite palliative measures progressively introduced by governments since the 1960s, continuing legal and illegal logging, unscrupulous commercially fostered forest encroachment, and traditional shifting cultivation practices have contributed to the further destruction of forest areas and denudation of grasslands. While the particular problems above are specific to Thailand, the country's plight is not unique. In the exploitation of natural resources, almost every country in the world faces the task of balancing, in its own time and way, the demands of economic and social development with environmental considerations.

The cumulative effect of environmental degradation was made dramatically and unfortunately clear in 1989, in a large-scale disaster in Southern Thailand in which torrential rain and flash floods on denuded land caused massive landslides and mud flows destroying villages and killing and injuring scores of people. Following that, to allow the government time to formulate and implement practical solutions, all logging of timber even in concession areas was prohibited by the Forest Reserve Act in 1990. Other wide-ranging consequences of this legislation aside, even purely in terms of elephants, the effects are complex. Certainly, elephants in the wild stand to benefit directly. However, placed at a serious disadvantage are the most 'visible' elephants in Thailand, the working elephants born and bred in captivity.

At the turn of the twentieth century, the numbers of working elephants in Thailand were estimated at some tens of thousands, of which a proportion may have been 'migrant workers' from neighbouring countries. Recent optimistic estimates place Thai elephants in captivity at around 1,500 (Sapana Sakya, 1994). Of these, some one hundred are government-owned, the rest being in private hands. This dramatic decline in their numbers is attributable to a variety of complex interconnected historical, economic, and social conditions, the enumeration of which is beyond the scope of this book. However, it should be mentioned that the rapid development of Thailand in the last decades of the twentieth century, while reaping many undoubted benefits, has also incurred environmental and social ills. Both benefits and ills have contributed to the degradation of the rural environment and erosion of traditional ways of life, affecting even the age-long profession of elephant breeding and care in the countryside (Cheun Srisavast, 1986). Additionally, it seems that the suspension of logging has contributed to substantial loss of livelihood of working elephants and their owners. Many have entered the already crowded tourist and entertainment industries, carrying tourists on jungle treks, or displaying their skills at special forest camps. Others eke out an increasingly peripatetic and precarious

Fig. 6
Modes of transport, past and present: mahout, elephant, and motor cycle at Surin. (Photograph Rita Ringis)

existence, offering rides at beach resorts, or wandering crowded city streets (Fig. 6).

In fact, epitomizing the plight of the dispossessed elephant, and arousing nation-wide publicity and compassion in late 1993 and early 1994, was the sad saga of Honey, a circus elephant calf. Seriously injured in a traffic accident up-country, Honey was brought by her owners to the care of the Dusit Zoo in Bangkok, where veterinarians could do little more than make her last days as painless as possible. According to the veterinarian, Dr Alongkorn Mahanopp, Honey was in fact only one of many elephant calves, some thirty a year, injured or ill, which are brought to the Dusit Zoo, unfortunately too late for anything but terminal care (Sapana Sakya, 1994).

Recognition of this case as a symbol for some of the current dangers besetting the Thai elephant prompted the Sueb Nakhasathien Foundation to convene a seminar at the Forestry Faculty at Kasetsart University, bringing together concerned professionals, government officials, and representatives from non-governmental organizations to discuss 'The Future of the Thai Elephant—Lessons Learnt from Honey'. Their deliberations were highlighted in a report in the *Bangkok Post* (Sapana Sakya, 1994), which has a distinguished record of bringing conservation issues to the public eye.

According to the report, not only accident injuries but also malnutrition from poor and unbalanced diets stemming from a depleted environment regularly bring domestic elephants into institutional care. Dr Pricha Puangkham, Veterinarian and Project Head of the recently established Thai Elephant Conservation Project of the Friends of the Elephant Foundation in the north of Thailand, confirmed that some ninety elephants were in the care of the Project, many as a result of overwork and injury, brought in by owners unable to cope with their elephants' terminal illnesses. Others, less lucky, are turned loose in the forest, abandoned, essentially, to perish. The seminar commended action being taken to deal with these increasing problems of elephant health and care (a vastly expensive proposition for indigent owners), in the projected establishment of two new elephant hospitals, one at Khao Khiew Open Zoo near Chonburi in 1996, and one planned by the Royal Thai Forest Department in the Ngao district at Lampang.

Also raised in discussions were serious doubts as to the current effectiveness of overlapping legislation concerning the protection of elephants. On the one hand, elephants in the wild are the responsibility of the Forest Department, under the Wild Elephants Act of 1921 and the later Conservation and Protection of Wildlife Animals legislation which provide some measure of sanctions against their illegal capture and exploitation. However, on the other hand, elephants born in captivity are considered the responsibility of the Agriculture Department and come under separate legislation, the Draught Animals Act, as beasts of burden, along with horses, cattle, and donkeys. While this separation of responsibilities was appropriate in the past, under present conditions it appears to work against the best interests of both types of elephants. For example, the registration process of domestic elephants is open to manipulation by the unscrupulous, allowing illegally captured wild elephants to be registered at the age of eight as elephants 'born in captivity' and therefore domestic. Additionally, there are suggestions that the identity papers of deceased but originally legitimately registered domestic elephants are 'transferred' to cover illegally captured wild elephants. Although this type of activity may yet be relatively minor, detection of its very existence points to a growing and undesirable situation for which legal sanctions are difficult to initiate let alone implement. Short-term solutions to this raised at the seminar took the form of suggested amendments to the legislation, which would bring all elephants, wild and domestic, under the protection of a single goverment department.

However, given that 'not even the experts at the seminar could give exact numbers for [wild] elephants or where they are found' (Sapana Sakya, 1994), it would seem that current investigations into providing solutions to the problems of maintaining and encouraging viable wild elephant populations in Thailand are as yet in their infancy. As assessed by Mattana Srikrachang of the Wildlife Conservation Division of the Royal Thai Forest Department, in the whole of Thailand today, there are some forty scattered areas where elephants persist, and according to the earlier quoted report of the AESG, few if any of these can support viable elephant populations in the foreseeable future.

While some headway in assessment is being made by government researchers of the Wildlife Conservation Division of the Royal Thai Forest Department, and the non-governmental Wildlife Fund Thailand, participants at the seminar were in agreement that extensive systematic research and the conducting of a proper census of elephants nation-wide were essential prerequisites for long-term remedies to this slow but steadily continuing devastation of the nation's heritage.

However, the future of the Thai elephant is not merely in the hands of diligent scientific researchers and the results that they may in time accumulate. For any results to bear fruit, sustained co-operation and understanding at all levels of the society is needed, from the humblest to the most exalted and powerful levels, to ensure the continuation of the elephant not only as a symbol of the country's historic past but as a vigorous participant in Thailand's dynamic progress into the future.

2 Elephants in Thailand

MILLIONS of years ago, before the advent of man, the distant ancestors of the modern elephant roamed the earth, and were prolific in regions of what is now North America, Europe, parts of the Middle East, Africa, India, and Asia, including China. Age-long processes of evolutionary differentiation, climate changes of the earth, and the coming of man the hunter resulted in the gradual disappearance of most of these proboscidean creatures.

However, closer to our own times, the widespread distribution of some of the nearer relatives of present-day elephants is documented in archaeological finds on almost every continent. For example, bone and ivory fragments, as well as cave paintings in Europe, attest to their one-time presence there. Carved ivory panels found in prehistoric Egypt and in the later Middle East indicate extensive trade of such items as luxury goods in those regions by the end of the second millennium BC (*Larousse Encyclopedia*, 1984: 201–5). The diversity of the finds suggests an evolution of numerous traditions ascribing a ritual and ceremonial role to the elephant in various early societies. As long ago as 2,500 BC, in the Indus Valley civilization at Mohenjo-Daro and Harappa, on the north-west Indian subcontinent, elephants were evidently tamed and honoured, their likenesses (as well as those of sacred bulls) engraved on intaglio seals which were possibly votive offerings. There are also suggestions that these numerous pictorial records may be indications of a yet undeciphered form of script.

In China, excavations of the ancient Shang Dynasty (1750–1045 BC) royal tombs have yielded bronze ceremonial vessels in the shape of elephants, as well as ritual objects carved from ivory. One tomb, evidently that of an animal lover, even included 'the real thing'—the noble's pet elephant to accompany him on his journey into the afterlife (Sullivan, 1984: 18). In fact, since recorded history, the elephant has clearly served man well: its ivory carved to enhance the delights of power and peace, its strength harnessed to promote commerce, and, unfortunately, to further refine the arts of war.

Today, of the hundreds of variations of the elephant's ancestors, only two species of the elephant family survive. These differ subtly in appearance and are generally designated by their geographical distribution: the African elephant (*Loxodonta africana*) and the Asian (sometimes called Indian) elephant (*Elephas maximus*). This

11

latter species, with subspecies named after or defined by the countries or areas in which it occurs, is increasingly threatened with extinction and is found in dwindling numbers in India, Sri Lanka, Nepal, Bangladesh, Bhutan, Burma, China, Vietnam, Laos, Cambodia, Malaysia, parts of Indonesia (Sumatra), and Thailand (Fig. 7).

The distinctive appearance of the Asian elephant, as inscribed on numerous ancient coins, suggests that along with some African elephants, the forefathers of present-day Asian elephants had found their way into the battle lines of Alexander the Great as well as into the elephant squadrons of Hannibal of the Carthaginians. The Roman historian Livy, in *The War with Hannibal*, recorded his admiration of the elephant's prowess while documenting in detail the beast's considerable deficiencies in battle. But the point had been made: this enormous creature was amenable to training. Indeed, in India the elephant had long been extensively trained to do man's bidding. However, no doubt astonishing to the classical world was the observation by Megasthenes in the fourth century BC that in India the elephant was also revered as of divine origin and thus considered godlike, particularly in its albino form (Giles, 1930: 65).

In fact, in India, to propitiate the vast and unpredictable forces of nature, men worshipped and placated gods who were believed to personify these forces. In time, a powerful Trinity of Gods, Brahma, Vishnu, and Siva, came to be venerated as representing the eternally recurring cycles of the forces of creation, preservation, and destruction in the universe. Other gods emanating from the earliest mists of time were also honoured: Agni, the God of Fire, and Indra, armed with his thunderbolt, the Lord of Rain and

Fig. 7
Thai elephants in a forest camp.
(Photograph courtesy of the Forest Industry Organization)

12

Storms, the destroyer of drought and thus the bestower of abundance in nature. To accomplish this vast task, Indra traversed the universe mounted on his celestial white elephant, Airavata, or Erawan as it was later known in Thailand (Plate 3). Thus from earliest times the elephant was associated with the blessings of rain, and nature's regeneration.

When Indian religious traditions and beliefs found their way, nearly 2,000 years ago to regions of South-East Asia, specifically to the region known today as Thailand, they were absorbed and adapted, over succeeding centuries, to suit local preferences and conditions (see Fig. 98). In time, the inherited traditions and mythologies, both Hindu and Buddhist, provided the philosophical and ceremonial basis of complex societies. Such traditions permeated necessarily diverse aspects of these societies, including even the rituals of capturing and taming elephants in the wild. Thus in those early centuries of the first millennium, Chinese travellers to now vanished South-East Asian kingdoms noted that elephants were indigenous and abundant in those parts, and that they were particularly prized there, being apparently docile and well suited to the then predominantly jungle terrain, as beasts of burden, and as vehicles of everyday and royal transport (Coedès, 1968: 51).

In effect, these observations of long ago indicate that ancient Indian customs of capture and training of elephants, as well as religious beliefs associated with them, had been imported and grafted on to indigenous, and apparently animistic beliefs and practices of the early inhabitants of the region. This is abundantly clear from the fact that until late in the nineteenth century and enduring into the early twentieth century in the large-scale hunting and training of elephants in Thailand, ancient forms of the Hindu gods as well as later Buddhist guardians were still invoked for the protection of the hunters and the hunted. Additionally, words and phrases of recognizably Indian origin were still customary in the training and directing of elephants (Giles, 1930a: 69).

From about the thirteenth century, a Thai people who were already followers of Buddhism gained sovereignty over the geographical area known in the present day as Thailand. There they established prosperous and influential kingdoms, and over the next six centuries or so, time and again repelled invaders, as well as asserting their sovereignty over neighbouring vassal states. In this they were greatly assisted by their strategic use of elephants. In fact, records in the Thai language dating from the late thirteenth century and documenting life in the Thai kingdom of Sukhothai refer not only to the hunting of elephants but also to the elephant's important position as an object of wealth and trade, as a help in war, and as a participant in royal and religious ritual (Griswold and Prasert na Nagara, 1971). The continuing importance accorded the elephant is clear from the fact that since the fifteenth century until the late nineteenth century, the Royal Elephant Department (Krom Kochabal) played a vital role in the

13

courts of the kings, and was responsible for the supervision of the hunting, capture, and training of elephants for the defence of the realm.

From about the mid-sixteenth century onwards, the Siamese cautiously welcomed other 'invaders', the Europeans from the West, lured there by adventure and trade in exotic goods, including trade in elephants. By the end of the seventeenth century, the capital of Siam, Ayutthaya, was a splendid city, vividly described by contemporary European traders in their detailed memoirs and reports (Fig. 8). These included records of royal processions with elephants (Plate 4), and the observation that 'elephants are esteemed the King of Siam's Principal Forces', numbering perhaps some 10,000. Such reports consistently documented the respect in which these beasts were held, and noted that they were never killed for sport (La Loubère, 1693: 89).

It was at this time, during the reign of the great King Narai (1656–88), who loved to hunt and train elephants, that some of the earliest writings about elephants and their care in Siam appear to have been compiled. These recounted traditional Hindu-derived beliefs about the divine origin of elephants, and enumerated the auspicious characteristics of elephants that ensured victory in battle, as well as those qualities that portended good or ill for their owners. Such traditions were no doubt based on the records of earlier centuries of observation of elephant appearance, behaviour, and 'character'. Compiled into manuals, and illustrated, these were guarded as secrets of defence. Of these now lost documents only relatively recent copies exist today in the form of beautifully inscribed and illustrated manuscripts, kept in the National Library in Bangkok (Fig. 9). In celebration of the sixtieth birthday of Her Majesty Queen Sirikit in August 1992, an early nineteenth-

Fig. 8
Built in imitation of Angkor Wat, these once gilded towers of Wat Chaiwattanaram are a reminder of the splendour of Ayutthaya in the seventeenth century. (Photograph John Ringis)

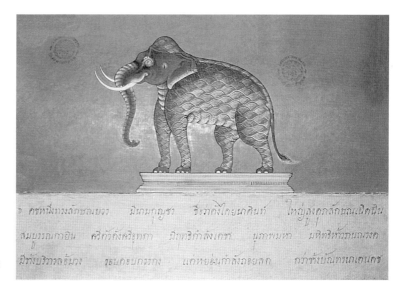

Fig. 9
Profile of an ocean-coloured elephant. The script praises the health and strength of this type of elephant, a natural leader. Early nineteenth-century manuscript. (Photograph courtesy of the National Library, Bangkok)

century version of these manuscripts was reproduced and published by the National Library to allow the general Thai public to enjoy this interesting elephantine heritage.

The relative rarity of Thai manuscripts on any subject, let alone elephants, is due to the fact that in 1767 the city of Ayutthaya, the great capital of nearly 400 years of Thai civilization, was destroyed, along with its monuments and documents after a long and bitter siege by the Burmese, then traditional enemies of the Thai. However, a new kingdom was quickly established and flourishing by the late eighteenth century in Bangkok, based on the revitalized traditions of Ayutthaya (Fig. 10). These traditions were challenged from the mid-nineteenth century onwards, with the second major

Fig. 10
Ancient and modern traditions of architecture coexist in Bangkok. (Photograph John Ringis)

influx of Europeans and Western ideas as Siam reopened itself to trade and diplomatic relations with the West.

From the late nineteenth to the mid-twentieth century, when some 70 per cent of the land was still forest, a major export item from Siam was teak wood, destined for the growing European middle classes. Numerous teak concessions were granted to Bangkok-based European trading firms such as the East Asiatic Company, the Bombay Burmah (later Borneo) Company, and Louis Leonowens and Sons, a company founded by the son of Anna Leonowens, the English governess known for her memoirs of her life in Siam. During that time, many Europeans followed the cry of 'Go East, young man!' and the forest jungles of Siam became home to a new breed, the foreign 'teak wallahs'. These men, with the help of local forest labourers, as well as hundreds of elephants and their mahouts, oversaw the meticulous classification and back-breaking work of extracting teak logs from the northern forests, guiding and supervising the journeys of the logs along the riverways of the kingdom until these reached Bangkok, ready for export some five years later (Fig. 11).

To regulate both the practices of the forest trade, and the welfare of the elephants working within it, the Royal Thai Forest Department was established, and a Law Concerning the Registration of Elephants was implemented by the early twentieth century. Then, in 1921, the promulgation of the Law for the Conservation of Elephants, possibly one of the earliest of such laws in the world, formalized many Thai traditional practices concerning elephants, including providing, in theory at least, for their conservation. Under this law (and by age-old tradition), elephants were not considered game animals, thus fines and penalties for killing or even wounding wild elephants were imposed (three years' imprisonment for killing and eighteen months for wounding). However, their capture was permitted, under licence,

Fig. 11
Harnessed elephants hauling logs.
(Photograph courtesy of the Tourism Authority of Thailand)

for use in logging and transportation. One in five elephants captured had to be given to the government. Export of elephants was prohibited except in special cases under special licence (Boonsong Legakul and McNeely, 1977). This is still the case today. One famous Thai elephant export of recent times was a handsome tusker sent to Sri Lanka to become the privileged bearer of the Buddha Tooth Relic in the annual Perahera Festival in Kandy. Another earlier and very influential export was welcomed to great acclaim in 1953 by the then Portland Zoological Gardens in Oregon (now known as the Metro Washington Park Zoo). This was a four-year-old Thai-born elephant called Rosy, whose arrival contributed greatly in time to the initiation of a continuing and successful programme of breeding elephants in captivity in the United States (Metro Washington Park Zoo, pers. com.).

The law also stipulated that any tame or captured wild albino elephants (*chang pheuak*), or elephants having other specifically outlined auspicious characteristics (*chang samkhan*), were to be presented to the Crown, and their captors or owners accordingly rewarded (Fig. 12). These characteristics, based on those recorded in the earlier mentioned illustrated manuscripts concerning traditional knowledge and observations about elephants, special or otherwise, are outlined in Chapters 3 and 5. While many of these traditional beliefs about elephants may seem interesting but basically fanciful to the European way of thinking, they are in fact thoroughly imbued in the Thai consciousness through custom, language, literature, religion, and art.

The best way to experience this at first hand, though not necessarily to understand it thoroughly, is to take a walk through the seemingly least traditional place in Thailand—Bangkok, today the home of the traffic jam *par excellence*. Here, in this highly urbanized concrete and glass jungle, the degree to which ancient traditions still hold sway in the Thai imagination is evident in the way the elephant, in any of its many forms, is everywhere present—to those who know where to look.

Wherever one may go, representations and images of the elephant in one form or another capture the eye. Not far from the Grand Palace area, presiding from a high gable of the modern National Theatre is a portly, seated, many-armed stone figure with the head of an elephant wearing a crown. Below, in a courtyard of the prestigious National School of Dance and Dramatic Arts, a larger but similar elephantine figure sits in solemn state on a raised dais, decorated with daily offerings of fresh flower garlands,[1] platters of food and drinks, and the ever-present sticks of burning

Fig. 12
Chang samkhan (auspiciously significant elephants) out for a stroll near the Royal Elephant Stables in the grounds of Chitralada Palace. (Photograph Rita Ringis)

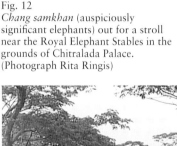

[1] Who garlands this statue in homage to its powers? Anyone, from trainee dancer to director, who works within that area, as the image serves both as Hindu deity to be honoured, as well as representative of ancient pre-Hindu indigenous animistic beliefs that require propitiation of the spirit of the land or *phra phum* resident in that particular place.

Fig. 13
Statue of Ganesha in the grounds of the
National School of Dance and Dramatic
Arts, Bangkok, with Wat Phra Keow
Wang Na in the background.
(Photograph Rita Ringis)

incense (Plate 5; Fig. 13). This well-tended being represents the
son of the Hindu god Siva. Known in Thai as Phra Phiganet
(Vighanessuan), he is customarily known in English as Lord
Ganesha—the God of Wisdom and the Arts, the Remover
of Obstacles, and the symbol of the Fine Arts Department of
Thailand (Plates 5 and 6). Enter any shop, and as like as not, a
smaller version of this same god is sure to be displayed prominently from a little altar, to ward off obstacles that may hinder
financial success. At amulet markets around the country, smaller
effigies of this elephant god are purchased for personal use as
charms for success in business ventures, or a little extra help in
school examinations (see Fig. 97). Ganesha's powers and idio-
syncrasies are also honoured up-country, in little spirit shrines
heaped with fresh and fragrant daily offerings, by those whose
livelihood depends on their work with elephants.

While Ganesha may be represented as rather cute and cuddly, of
a different type entirely is another elephant, decidedly more
elegant though none the less anatomically imaginative. Jingle your
coins, and look at them carefully. Amongst the variety, you may

18

Fig. 14
Indra on the three-headed Erawan elephant above the entrance to the National Stadium, Bangkok, twentieth century. (Photograph Isabel Ringis)

Fig. 15
Emblem of Indra on Erawan at the Bangkok Metropolitan Administration Headquarters, twentieth century. (Photograph Rita Ringis)

be lucky to find an old coin that bears the image of a three-headed elephant. Attend a football match at the National Stadium, and while waiting for your entry ticket, observe the façade above you decorated with two handsome identical and symmetrically aligned Art Deco statues, each depicting a powerful warrior seated at the neck of a majestic three-headed elephant (Fig. 14). Board a crowded river-boat, and visit Wat Arun on the Chao Phraya river-bank, that towering temple whose silhouette on the Bangkok skyline is synonymous with Thailand for tourists. At each of the four cardinal points of this lofty tower is visible the same celestial being mounted on that same three-headed elephant (Plate 8).

In all cases, the celestial rider represents the originally Indian Hindu god Indra, mounted on his three-headed white elephant, Airavata, presiding over all from his celestial palace at the centre of the universe. However, the Buddhist Thai have thoroughly adapted and incorporated the potent symbol of Indra on his white elephant, known in Thai as Erawan, into their own traditional beliefs, even into the principles of town planning. In fact, the god Indra is considered a celestial guardian of Bangkok, as reflected in the opening phrases of the Thai name of this city, 'Krungthep Rattanakosin ...' which may be translated as 'City of Angels, Precious Abode of Indra ...'.

Indra's divine powers, together with the might of his white elephant, Erawan, are invoked even in the decidedly secular side of administering this great city. Thus, as you pause by the traffic-lights and marvel at the congestion of vehicles, jammed amongst them is likely to be an enormous yellow garbage truck, and stencilled on its doors, is a logo of the figure of Indra riding Erawan, in this case with six tusks. In fact, this motif derives from the official emblem of the Bangkok Metropolitan Administration, emblazoned on the modern building housing the headquarters of the Bangkok city government, and is based on an early twentieth-century European-influenced design by Prince Naris, the designer of the Marble Temple (Fig. 15). In turn, the Bangkok Metropolitan Administration building itself is understandably sited opposite the great Buddhist temple of Wat Suthat, built in the early nineteenth century and considered at that time as the sacred centre of the then burgeoning city of Bangkok (Fig. 16). At Wat Suthat, the high and glittering pediments or gable-boards enshrine the same graceful figure of Indra on his sacred elephant, Erawan, rendered in the style of the nineteenth century (Plate 9). Reaffirming the continuing relevance of these ancient symbols is a spectacular pediment of Indra on Erawan, created in the present reign at Wat Rajburana near the Chao Phraya River (Plate 10). From this celestial elephant, revered originally in ancient India, and certainly respected still today in Thailand, derives the concept of the sanctity of the earthly white elephant.

Perhaps the best known, and at the same time most confusing place to come across some elephant images of the more ordinary kind is at the Erawan Shrine, a more recent 'centre' of the city, in

Fig. 16
Wat Suthat, with the Giant Swing in the foreground, opposite the Bangkok Metropolitan Administration Building. (Photograph Rita Ringis)

a business and booming shopping area, situated at the major inter-section of Rajadamri and Ploenchit Roads (Fig. 17). There, to the throb of passing traffic and plangent strains of Thai traditional song and music, in a cloud of fragrant incense and stifling traffic fumes, hundreds of devotees pray every day for good fortune. This shrine is famous for its beneficent powers throughout Thailand and South-East Asia (where it is almost universally but incorrectly known as the Four-faced Buddha Shrine). In return for wishes granted and hopes fulfilled, the supplicants hire dancers and musi-cians, and daily offer the shrine's presiding deity, the four-faced and many-armed Hindu god Brahma, literally tons of multi-coloured garlands, food (salvers of pigs' heads and elegantly arranged boiled eggs being popular). Most notable, and certainly unavoidable at the shrine, are numerous wooden elephants, many of which are almost life-sized, brightly garlanded and covered with gold leaf (Fig. 18). Charitable organizations benefit from the many large and small financial donations customarily made at the shrine.

This shrine was actually built as the protective spirit house or *sarn phra phum* for the original and much-loved government-owned Erawan Hotel which stood here for decades before being replaced in the early 1990s by a more luxurious international hotel. The shrine is a *mélange* of animist and transmuted Hindu beliefs, being a place of propitiation of the spirit of the area as well as a homage to the god Brahma, but known after the name of

Fig. 17
Dancers at the Erawan Shrine.
(Photograph Eileen Deeley)

Fig. 18
Supplicant and gilded wooden elephants
at the Erawan Shrine. (Photograph
Rita Ringis)

the original hotel, Erawan, after the elephant of Indra. In fact, the numerous wooden elephants presented there are in honour of this celestial elephant. While the Thai are clear as to who-is-who here, it is no wonder there is confusion elsewhere as to its central figure of honour.[2]

Away from all this colourful confusion, visit the Temple of the Emerald Buddha[3] and stroll through its beautiful courtyards housing ornately decorated Buddhist assembly halls and monuments. Join the other countless visitors before you, patting for good luck the now shiny forehead of a naturalistic bronze elephant replica, one of many arranged in formal groups within the complex (Fig. 19). These are memorials to the royal white elephants of the kings of the Chakri Dynasty which has reigned since the foundation of Bangkok in 1782. The royal white elephants are also memorialized in full-size bronze sculptures at the foot of

[2]Attesting to the widespread fame of this shrine's powers, taxi drivers from as far afield as Singapore display a miniature effigy of the Brahma on their dashboard for good luck and swear that it is the 'Four-faced Buddha of Bangkok' of the Erawan Shrine.

[3]Popularly known as Wat Phra Keow, the temple's formal Thai name is Wat Phra Si Rattanasatsadaram.

Fig. 19
Memorial with bronze replicas of white
elephants of the Chakri Dynasty at the
Phra Si Rattana Chedi, Temple of the
Emerald Buddha, Bangkok.
(Photograph Rita Ringis)

Fig. 20
A soldier and memorial bronze white
elephant on guard at the stairway to the
Chakri Maha Prasat Throne Hall at the
Grand Palace, Bangkok. (Photograph
Rita Ringis)

the imposing staircase at the entrance to the Grand Palace, a late
nineteenth-century structure, which was built on the original
site of the white elephant stables of the early Chakri kings of
Bangkok (Fig. 20).

Enter the Chapel of the Emerald Buddha, where the most venerated Buddha image in Thailand presides over humbly seated devotees and tourists from a lofty, gilded, and canopied pedestal
throne, surrounded by numerous golden images, some of them
crowned and in sumptuous regal attire. In contrast to the scintilla-
tion of the mosaic mirror decorations and the bright glow of the
gold of the 'altar' area, the surrounding walls, covered from floor
to ceiling with painted episodes from the Buddhist scriptures, may
at first glance seem muted, but as the eyes grow accustomed, the
brilliant colours of the murals begin to emerge, and a multitude of
miniature scenes becomes evident. On the eastern wall facing the
presiding images, a vast mural of almost life-sized figures stretches

22

Fig. 21
Calling the Earth to Witness and
Victory over Mara. Mural at Wat
Dussidaram, Thonburi, nineteenth
century. (Photograph Eileen Deeley)

to the ceiling, depicting a scene of dramatic turmoil, dominated by
two great war elephants apparently facing each other. To the
devotee, this is instantly recognizable as the moment of the
Enlightenment of the Buddha, Calling the Earth to Witness, and
attaining his Victory over Mara. This episode, almost symmetrical
in composition but not in detail, is traditionally rendered in this
manner on the eastern walls of temples all over the country
(Fig. 21).

Highlighted against a vivid red background, in the upper right-
hand corner of the scene (from the observer's point of view)
looms a magnificent tuskered elephant caprisoned for war,
bearing the crowned demon Mara, Lord of Worldly Desire and
Delusion, leading his army of demons and monsters, soldiers and
men, clad in Chinese, Arab, Persian, Indian, and European fashions of centuries ago. All bristle with arms, brandishing cudgels,
pikes, swords, and flintlock rifles; even the elephant flourishes a
sword with his trunk. The quarry of this frenzied assault is
depicted high above, at the centre of the mural, as if in the still
eye of the storm: the red-robed Gautama seated under the bodhi
tree, his right hand resting on his right knee, his fingers touching
the earth. Directly below, dominating the central panel of the
mural, is the sinuous earth goddess, Thoranee, whom Gautama
has thus summoned to witness his virtues. Wringing out her long

23

tresses, Thoranee unleashes a cleansing flood, washing away the demons of doubt and delusion. Their convulsed struggle against raging waves and voracious aquatic monsters (Plate 11; see also Fig. 64) is depicted on the viewer's left. High above at the upper left-hand side of the mural, again silhouetted in glowing red, in almost mirror image of his aggressive form on the right, is the now subdued Mara, still mounted on his war elephant, proferring lotus flowers instead of weapons, paying homage to the victorious Gautama, henceforth called the Buddha or Enlightened One. This scene, of utmost importance in the iconography of Thai Buddhism, has something in it for everyone, from the literal pictorial representation of the allegory of one of the deepest noble truths of Buddhism, to the humorous focus on the often bawdy aspects of the demons' and warriors' anatomies, as well as the superbly detailed documentation of magnificent elephants, so dear to the Thai heart.

Ironically, today's travellers are likely to meet their first 'real' Thai elephant in a most unexpected way. Not unusual nowadays is a startling sight: the occasional elephant, large and dusty, ridden by his weary mahout, and towering above the trucks and buses of Bangkok's traffic-congested streets, or even loitering in the back lanes (Fig. 22). The elephant and his rider will undoubtedly have travelled from the north-eastern province of Surin, traditionally famous for its elephants and their expert keepers. For some years now, during the dry season, the hottest part of the year, many such elephants and their keepers make the twenty-day journey to

Fig. 22
A pair of pachyderms loitering in a Bangkok lane. (Photograph Kim Retka)

24

Fig. 23
Ducking under the elephant for good luck. (Photograph Kim Retka)

the city in search of food, and a little extra money. Moving sure-footedly in and out of what passes for traffic lanes, and more often than not going against the flow of traffic, this apparently placid and long-suffering beast provides a welcome diversion for even the most jaded of Bangkok's commuters. Painted on the elephant's flank is an invitation that for the traditionally minded is hard to resist. Good luck, particularly for women of childbearing age, is ensured by a small donation, and the courage to walk under the belly of the elephant (Fig. 23). Most customers for this service will be found in the small but crowded lanes off the main roads, on the many construction sites employing up-country workers in Bangkok's overwhelming building boom. Here and there, as the elephant plods between construction sites, sympathetic householders or their maids will provide water and bananas for the weary animal. In the lanes or *soi*, any grassed area will become an immediate and welcome buffet. This truly memorable sight of a jumbo country bumpkin in the concrete jungle is a symptom of the elephant's plight in Thailand, no longer a lord of the realm nor a helpmate in labour, but a dispossessed mendicant, dependant on hand-outs from strangers.

Throughout much of the twentieth century, the domesticated elephants have laboured in many undertakings. These have included a major role in road construction in the mountainous terrain of the north during the 1970s. However, the majority of elephants were employed on a seasonal basis in the timber industry until the logging ban in 1990. Rapid social and economic change in the last decades of the twentieth century have led to a decline in some rural traditions, affecting many elephant owners' customary livelihoods (Cheun Srisavast, 1986). Thus the owners and their elephants must increasingly depend on new ways of making a living. Both in the north and north-east of Thailand, this has meant entering the entertainment industry, hiring elephants out for shows, movies, festivals, and to the tourist circuit. As Thailand is a major international tourist destination, there are many places where elephants are employed, going through their paces for local and international tourists alike.

Training camps, both commercial and government, may be visited in parts of northern Thailand (Fig. 24). These allow tourists to see the elephants in spectacular natural surroundings of forest and river, going about their tasks under the direction of their mahouts (see Plates 1 and 2). Demonstrating their uncanny obedience to their mahouts' commands, the elephants perform various complex activities. Working singly or in tandem, the elephants display their skills in lifting and placing logs in precise and balanced order. In harness, they pull well-worn teak logs to display their strength and co-ordination. They bathe, and are bathed and scrubbed by their masters, with both elephants and mahouts obviously enjoying the whole process (see Fig. 41). These spectacular ablutions are not merely a ploy to entertain tourists,

Fig. 24
Elephants and mahouts perform for tourists at a forest camp. (Photograph Monique Heitmann)

but are essential for the continuing well-being of the elephant, contributing to the healthy and sore-free malleability of the skin, and maintaining its body temperature at acceptable levels, for unable to sweat, an elephant is particularly vulnerable to the heat. Throughout all of these activities, the close and responsive relationship between elephant and keeper is clearly evident.

At the end of the show, the tourists are invited to experience a brief ride, under the control of the mahout, on elephant-back through the jungle. Purists may decry this commercialization of noble beasts. However, these so-called tourist 'traps' must represent Elephant Heaven, for at no stage do the elephants perform 'tricks' or activities that are unnatural or demeaning.[4]

Perhaps the most spectacular way to meet a multitude of elephants and view their many attributes and skills is at the Elephant Fair and Round-up in the north-eastern province of Surin,

[4]Compare this to many traditional zoos, which still exist, and where most people meet their first elephant. There the one or two resident chained elephants pace restlessly back and forth in the confined quarters of an 'elephant house' generally constructed to look like a poor man's Moghul palace. Or, for that matter, consider the traditional circus, where the elephants, of necessity confined for most of the time, are trained to balance awkwardly on tiny stands so that the audience may applaud such vast bulk being so ignominiously obedient to some tiny skimpily clad be-spangled 'trainer'. Thankfully, such 'traditional' places and activities are on the way out, and intelligent and humane methods of displaying animals and their capabilities in well-planned zoos are considerably contributing to people's understanding and appreciation of the richness of nature's infinite variety.

26

Fig. 25
Elephants in a battle enactment at the
Surin Elephant Round-up. (Photograph
Rita Ringis)

renowned for its ancient Khmer temples, its exquisitely crafted textiles and shimmering silks, and certainly not the least, for its elephants. Held annually on the second weekend in November, the Elephant Round-up, now entering its fourth decade, was established by the Surin provincial authorities to highlight the importance of the role of elephants in the history of the province (Fig. 25). While thus keeping alive the rich traditions associated with elephant care in the region, the fair also contributes to the present welfare of the animals and their keepers, as well as boosting the general economy of the province.

In former times, a round-up was the natural culmination of several months' long actual expeditions to the forests, to capture elephants and drive them to holding areas for training and 'conscription' as war elephants. Until the early twentieth century, the annual round-up was also a time of celebration for the local people, as well as a display for visiting dignitaries, presided over by royalty. The present Surin Round-up reveals in microcosm the former methods of elephant capture, the traditions of pageantry in which elephants participated, as well as the many skilful capacities and endearing attributes of this most intelligent of animals.

Today, overseas and local tourists alike flock to the Surin Round-up to relive, albeit in a somewhat artificial but certainly more democratic setting, some of the pageantry of the past (Plates 12 and 13). Seated in multi-tiered grandstands temporarily erected on either side of a vast field handsomely decorated to resemble a forested fort, the audience cannot help but marvel at this grand annual reunion of hundreds of domestic elephants from all over the country. For most of the onlookers, this may be the first time they have seen at such close range such large numbers of elephants moving *en masse* (Fig. 26). Immediately evident, and somewhat surprising, is an overwhelming impression of monumental cohesiveness of the animals. Though the elephants taking part in the show have been gathered from all around the country, together (each directed by its mahout) they move almost like one single gigantic many-legged organism, loping in a fluid surprisingly graceful motion whose uniformity is broken only here and there by the uneven trotting and lolloping of the elephant calves struggling to stay by their mothers' sides (see also Fig. 50).

In this orchestrated mass run-around, as in all the other exercises displaying the elephant's obedience, dexterity, and skill, most clearly evident is the responsiveness with which each animal reacts to its mahout's instructions. These are given by actual verbal commands, or more often, by pressure exerted by the mahout who straddles the elephant's neck, and uses his feet and toes to pressure the sensitive backs of the elephant's ears to convey his instructions (see Fig. 39). The mahout also carries an elephant goad, a stick with a sharp hooked point, which he may apply to the elephant's head, or other parts of the body. Although this looks cruel, and invariably arouses some distress in the viewer, properly applied, it does not hurt, but stimulates certain points traditionally held (or

Fig. 26
Elephants moving *en masse* at the Surin Elephant Round-up. (Photograph Rita Ringis)

created by training) to produce appropriate responses from the elephant.

After welcoming the province dignitaries in a suitably elephant-ine way (Fig. 27), during the Surin show, the elephants perform either individually or *en masse* various activities that illustrate their strength, nimbleness, and good judgement. In a hilarious tug of war with several scores of no doubt recent conscripts to the Thai army, the elephant wins, but only just, because kilogram for kilogram, unexpectedly, the elephant is weaker than man (Toke Gale, 1974: 4), that is, as a proportion of its body weight, the ele-phant can carry less than can a man or even horse. Thus, the 'trick' in such a tug of war against the elephant is in the calculated choice of the number of men and their combined weight. In such a contest, the onlookers' expectations of the outcome are over-whelmed by the sheer numbers of men, say some fifty, against one elephant. But only if the combined weight of the men is consider-ably less than that of the elephant will the elephant win. For their pains, the predictable losers in the test of strength, the hapless conscripts, are awarded the dubious pleasure of resting, and all obediently lie down on the grass arena in a straight line, side by side, each man separated by a mere smidgen of distance. Then to the hushed and terrified anticipation of the audience, the elephants demonstrate their capacity for extraordinary precision and care in working as a team. Responding to their mahouts' directions, the elephants queue, and in single file carefully negotiate the line of

Fig. 27
Elephants and dignitaries at the Surin Elephant Round-up. (Photograph Rita Ringis)

Fig. 28
Elephants performing at the Surin
Round-up: negotiating recumbent
recruits. (Photograph Rita Ringis)

recumbent recruits, delicately placing their feet into the small widths of space between each vulnerable man (Fig. 28). At the safe conclusion of this remarkable exercise, both the crowd and the elephants trumpet their delight. The show ends with various grandly choreographed processions depicting kings duelling on war elephants, attended by scores of brightly costumed soldiers (Fig. 29). Also re-enacted are ancient religious and propitiatory ceremonies, with elephants, attendants, and musicians by the hundreds, ceremonies that in fact still take place today in seasonal festivals all over the country.

After the show is over, individual elephants may be hired for the now traditional ride on elephant-back into town, and tourists who may think twice about crossing a street in Bangkok throw caution to the winds, and queue patiently in line at the elephant mounts, high platforms reached by stepped or ladder-like structures beside which the elephant, with the mahout mounted at its neck, awaits its passengers (Figs. 30 and 31). From these platforms set slightly higher than the elephant's height, the intrepid tourist may negotiate himself and perhaps his companion confidently into the elephant chair or howdah (a box-like saddle) strapped on to the elephant's back with sturdy chains. Thus seated, and thankfully with the mahout in control, the tourist may, on this little journey, experience, in his imagination at least, travelling in the style of the nobility of former times.

Today the journey wends its swaying measured way through the streets of this provincial town, with the elephants bearing the sun-hatted, camera-toting tourists (shivering with delight at their

Fig. 29
Procession at the Surin Elephant Round-up. (Photograph Rita Ringis)

Fig. 30
Elephant awaiting a rider at a rustic elephant mount. (Photograph Monique Heitmann)

31

Fig. 31
Elephants and mahouts awaiting
tourists in the town of Surin.
(Photograph Rita Ringis)

Fig. 32
Annual traffic jam in the town of Surin
after the Elephant Round-up.
(Photograph Rita Ringis)

daring). The elephants plod sedately side by side with passing cars,
overcrowded buses, zigzagging motor cycles, and amazingly non-
chalant pedestrians (Fig. 32). To cheer the procession on, the local
merchants and their families are out in full force lining the streets.
However, the custom of offering delicacies to the elephants, particularly brimming buckets of cooling water into which that
sinuous hose of a trunk drops gratefully, may give the unaccus-
tomed riders some momentary disquiet.

The elephants of Thailand—both the wild and free, and the
tame and domestic—are an integral part of the nation's heritage,
historical, cultural, and natural. Whether the elephants continue
to flourish and thrive in viable populations or continue to decline
into mere tokens of a theatrically re-created splendid past is in the
hands of present generations.

3   The Thai Elephant from Trunk to Tail

WHILE there are regional variations within the Asian or *Elephas maximus* species, since elephants in the South-East Asian continental regions recognize few politically determined boundaries, the characteristics and behaviour patterns discussed below are based on generally accepted observations about the Asian elephant, but with specific reference to the elephants of Thailand (Fig. 33). However, this survey is by its very nature limited to what might intrigue or appeal to the general interest reader; for detailed facts and comparative figures of the Thai elephant's vital statistics, various specialist works are listed in the Bibliography.

Since the elephant has played a vital role in all aspects of Thai life, many apparently non-scientific descriptions of its physical

Fig. 33
Elephants, trunk to tail, at the Surin Elephant Round-up. (Photograph Rita Ringis)

attributes and characteristics have been incorporated into long-standing oral traditions as well as periodically recorded in Thai treatises on elephant care, known as *Gajasastra* and *Kotchalakshana*. These are in the form of illustrated manuscripts, and at first hand their contents may appear to be far-fetched records of folk wisdom and poetic fantasy (see Fig. 9) But given the fact that the Thai people have had a conspicuous success in the capture, taming, and training of elephants for many centuries, and considering the undoubted importance of the elephant in the economic, military, religious, and social history of Thailand, many of the observations recorded in the manuscripts are surely largely based on reality and pragmatic practices (Fig. 34). However, the manuscripts and folk traditions, because of their very nature, are not strictly in accord with modern evolutionary theory, and understandably ascribe a divine origin to the elephant, based on a variety of beliefs, many of them inherited from ancient India. Thus, rather than following the pedestrian processes of evolution, one such belief proposes that four castes or families of elephants, the noble ancestors of modern elephants, were divinely transmuted from the petals of a golden lotus and its many stamens by four Hindu gods at the time of the creation of the universe (Giles, 1930a: 63–4; Fine Arts Department, 1990). The lotus itself had miraculously emanated from the navel of one of the gods, Vishnu (Phra Narai in Thai), as he reclined on the face of the waters of the inchoate universe, dreaming his Cosmic Dream of the universe-to-come (Fig. 35).

Interestingly, these traditional beliefs could be said to fore-shadow, somewhat poetically, evolutionary theory which recognizes that present-day elephants evolved from earlier but now vanished ancestors. Tradition and myth hold that in time, from the four castes of divine elephants thus created by the gods, many

Fig. 34
Profile of an elephant and its mahout at the Surin Elephant Round-up.
(Photograph Rita Ringis)

Fig. 35
Reclining Vishnu. Lintel at the east
entrance of the main sanctuary at Prasat
Khao Phnom Rung, Buriram Province,
tenth–thirteenth century. Inherent in
the symbolism of this scene is the (re-)
creation of the world and its inhabitants,
including elephants. At the upper right
and left, *hastilinga* birds hold limply
dangling elephants. (Photograph
Rita Ringis)

other types and families of everyday elephants were derived, and
that these elephants still retain distinct characteristics of their individual divinely ordained castes (Fig. 36; see also Plate 18). Just as
a rose is a rose is a rose, by the untutored eye, an elephant could
also thus be perceived. But the tutored eye, enhanced by the lens
of myth, discerns a rich variety which is briefly summarized below
(Giles, 1930a, 1930b; Fine Arts Department, 1990).

Not surprisingly, Thai elephants are classified according to their
possession of a variety of qualities originally imbued in the four
main castes of their divine elephant ancestors. The first and
highest caste, the Brahmin, derives from the petals transmuted by
the god Brahma, and this group is known in Thai as Phromapong
(the family of Phra Phrom or Brahma). Ownership of such elephants bestows longevity and wisdom. Elephants transmuted by
the god Siva belong to the Kshatriya caste of monarchs and

Fig. 36
Multi-hued elephants of myth. Detail of
mural on column at Wat Suthat,
nineteenth century. (Photograph
Isabel Ringis)

35

warriors, and in Thai are known as Issuanapong (the family of Phra Issuan, or Siva). They impart wealth and power to their owners. The Vaisya or commercial and agricultural caste of elephants, ensuring a good rainfall and thus prosperity, are from the Vissanupong family, deriving from lotus stamens transmuted by the god Vishnu (Phra Narai in Thai), and like the god himself, they radiate the qualities of mercy and love. The fourth caste of elephants, the Sudra or servile, in which both evil and good qualities fluctuate, derives from the God of Fire, Agni (Phra Phleung in Thai), and are called Akkanipong. Owners of these elephants are assured material pleasures and plenty. Each caste or elephant family is also described in terms of distinctive physical characteristics such as size, body shape, colour, and skin texture.

From within these castes, elephants of unusual or rare appearance are classified as important and auspicious elephants (*chang samkhan*), and by law are required to be given to the Crown. These include elephants of highly unusual colouring (*chang si pralart*), black-coloured elephants (*chang niem*), and most rare and thus most important and auspicious of all, the white elephant (*chang pheuak*). Supremely valued above all elephants, the white or albino elephant may be born from any of the above four family types, and is regarded as the earthly manifestation of the Celestial White Elephant, Erawan (Plates 14 and 15). Possession of such an elephant by a king is believed to be a sign of the king's great virtue and power. The characteristics of white elephants are discussed in Chapter 5.

While traditional writings enumerate by name and auspicious characteristics many mythical elephants whose powers have brought victory in war and prosperity in peace, 'undesirable' elephants are also described and depicted in the manuscripts, frequently in amusing detail. Thus owners are warned against the pig-snouted, hump-backed, big-footed, and ragged-eared elephants, as well as those with various other characteristics that depart from the desirable norm (Fig. 37).

However, the modern elephant, whether African or Asian, though perhaps not of such picturesque divine origin is none the less a miracle of evolutionary processes: it could be said to be an inspired exercise in balance and counterbalance, of prodigious strength allied to delicate precision of movement, at once hampered by its bulk, yet capable of forceful action and unexpected buoyant grace. This combination of unlikely qualities is reflected in the elephant's physical appearance, and recognition of the incongruity no doubt was the source of an ancient folk-tale in India. Six blind men, after individually touching an elephant, described it thus: it is like a snake (the trunk); it is like a knife or a sword (the tusk); it is like a fan (the ear); a rope (the tail); a pillar (the leg); a wall (the body and skin). The same idea is expressed in a Thai proverb, 'Taa boht khlam chang' ([Like] a blind man feeling an elephant), which is used to describe someone ignorant of all the facts of a given situation.

Fig. 37
A thoroughly undesirable ripple-backed elephant. Nineteenth-century *khoi* paper manuscript. (Photograph courtesy of the National Library, Bangkok)

While the African and Asian species have distinct differences, some of which are not immediately apparent, they are recognizably of the same family, and share certain common characteristics. Effectively limiting the elephant's vision are the comparatively tiny eyes set in the large head. This itself is set on a 'too short' neck, thus further limiting the creature's peripheral vision as any extensive turning of the head to see what lies behind is constrained by its anatomy. However, there are compensations. The head, though housing the largest brain of any land mammal, is not as heavy as it looks. Its interior is not all solid bone but a honeycomb-like structure, thus making the head light enough for manoeuvrability, but also sufficiently large to provide balance and attachments for the extensive system of muscles supporting the tusks and trunk (Figs. 38 and 39) (Boonsong Legakul and McNeely, 1977: 635).

Fossil studies of the elephant's earlier ancestors indicate that the trunk evolved from the fusion of a type of upper lip and nose, which had originally developed as a balance to an even earlier rather long jaw with lower tusks used for digging. Subsequent evolution led to the loss of the lower tusks, the reduction of the lower jaw into a spout-like chin structure, common only to man and the elephant (see Fig. 34), topped by the now elongated trunk (Boonsong Legakul and McNeely, 1977: 635). The modern elephant's tusks, weapons of defence or tools for digging up roots, also evolved from the elongation of the elephant's ancestors' upper incisors or front cutting teeth, not, as is frequently falsely assumed, from the elongation of its ancestors' canine teeth. Elephants have no canines.

For chewing of its high-fibre food, the modern elephant has developed a unique and enviable system of dentition, of tooth replacement some six times within its lifespan. While its cheek teeth premolars are present at birth, these are shed and replaced at the end of the first year, the sixth year, and the ninth year.

Fig. 38
A mature Thai tusker at the Lampang Elephant Training School. (Photograph Rita Ringis)

Fig. 39
A young tusker and its mahout at Surin. (Photograph Rita Ringis)

Following that, as the sets of molars are worn down through use, they are crowded out and gradually replaced by new sets growing simultaneously (almost 'escalator-like') under the old sets in the jaw (Boonsong Legakul and McNeely, 1977: 635) Though replaced thus frequently, it is, incongruously, the state of the teeth that largely determines the elephant's lifespan, whether African or Asian. While romantic myths in the past suggested lifespans of over 100 years, of venerable ancients commanding herds by the power of their mysterious antiquity, the sad truth appears to be that once the teeth wear down and are no longer replaced, the elephant is a captive of his own anatomy. Recent research in Africa and common knowledge among 'elephant men' in Thailand indicate that poor teeth lead to poor nutrition, increasing debility and death either by starvation or other means compounded and exacerbated by weakness.

Though they share certain characteristics as related above, the African and Asian elephants differ markedly. While the Asian elephant has been hunted and tamed for thousands of years, and has worked as a beast of burden for man, as a warrior in battle, and as a stately bearer of kings and nobles, the African elephant has on the whole not responded to taming, except on a very minor and occasional scale. As the brain size of the Asian elephant is said to be larger, various suppositions about its superior intelligence have been made. It would seem that considering present lack of comparable scientific evidence, subjective judgements enter here: whether being a 'slave' to man overrules living free, but being slaughtered on a large scale for meat and ivory.

Having evolved in different climates and circumstances, African and Asian elephants also differ in terms of their physical appear-

38

ance. Generally speaking, the Asian elephant, on average weighing about 3–4 tons and standing at a height of around 2.5 metres, is smaller in size and lighter in weight than its African cousin, which may average a weight of 6–7 tons, and reach a height of approximately 3.4 metres.

In profile, the African elephant clearly displays a dip in the spinal ridge between the fore and hind quarters, giving it a distinct waisted or saddle-back shape, whereas the Asian elephant has a generally convex, humped spinal ridge, curving gently downwards from the upper part of the spine (see Figs. 33 and 39). Thai treatises on desirable and undesirable characteristics of elephants are specific as to the advantages of the smoothly curving 'bent bow' shape of the spinal ridge. Elephants with spines that are not smooth but display major corrugations are to be particularly avoided, as they are said to bring misfortune to their owners. The logic is clear: a beast of burden would certainly be hampered by such a characteristic (see Fig. 37).

Unlike the elongated forehead and flat crown of the African elephant, the Asian elephant has a deeply lobed forehead and prominently twin-domed crown or head. Though both species share the apparent reflex action of constant flapping of the ears, the shape and size of the ears differ. While the African's ears are much larger, almost circular and set prominently rufflike at the head and neck, the Asian elephant's ears are set lower and closer to the side of the head, and are smaller, almost triangular in shape. The popular 1950s writer on elephants in Burma, J. H. Williams (known to his readers as 'Elephant Bill') likened these ears most appropriately to 'little maps of India'.

Both in Burma and in Thailand, the outline of the upper edge of the ear of the elephant is said to be like that of the banyan leaf. The simile of the leaf is apt, for just as a leaf curls at its outer edges with age, so the upper edge of the Asian elephant's ear 'curls' or folds over gradually at about its tenth year. According to U Toke Gale, in his fascinatingly detailed and exhaustive work, *Burmese Timber Elephant*, the increasing size of this fold is a useful and immediate indicator of the Asian elephant's approximate age, with every inch of fold representing some twenty years (see Fig. 34). While the condition of the teeth is also indicative of age, this method is understandably limited, particularly in the wild elephant.

In both African and Asian elephants, the constant flapping of the ears, while serving a useful purpose in keeping insects at bay, is said by some to be a cooling mechanism, by others a reflex action, similar to the blinking of the eyes in humans (Toke Gale, 1974). The regularity of this flapping action is also considered an indicator of the health of the elephant. The backs of the elephant's ears are prominently veined and thus extremely sensitive, and in the case of the Asian working elephant, they could be called its 'gearbox', for the rider or mahout, seated at the neck of the elephant, uses his toes and feet to stimulate these sensitive areas, directing and controlling the trained elephant's larger movements rather like pedalling a bicycle. This is particularly evident when a mahout is urging an elephant to run.

To compensate for its relatively poor sense of sight, the elephant's sense of hearing is acute, capable of differentiating sounds at a great distance, filtering out familiar non-threatening sounds, and reacting to sounds that may portend danger. While sudden unexpected sounds may cause panic in the Asian elephant, interestingly, certain seemingly harmless sounds, when repeated rhythmically for considerable periods of time, may also alarm the elephant, causing it to move away from the sound source, a fact well known and useful to the timber workers of the past in South-East Asia. Toke Gale (1974) cites instances of herds of wild elephants being diverted from their path of potential rampage by timber workers rhythmically thumping sticks on timber logs.

Sounds produced by elephants themselves vary from the ear-shattering trumpeting of rage or warning, to snorts of apparent delight and pleasure, through to low rumbles and throaty purrs. Traditional Thai writings describe some of these sounds (no doubt made by auspiciously desirable elephants) more poetically as 'voice like a trumpet' or 'whisper of the conch shell'.

Recent African studies cited in the *National Geographic* (Payne, 1989) suggest that elephant sounds may include forms of long-distance communication, many of them at frequencies inaudible to the human ear. Whether this is also the case with the Asian elephant has yet to be extensively scientifically documented. There

Fig. 40
Profile of a Thai tusker and its mahout.
(Photograph Monique Heitmann)

are suggestions that since the Asian elephant evolved in a relatively 'closed' tropical jungle climate, as against the wide open grasslands of the African elephant evolution, its olfactory sense rather than its sense of hearing may be more highly developed.

While both male and female African elephants invariably have long tusks, this is not the case with Asian elephants, where variations exist. For example, Sinhalese male elephants rarely have tusks, whereas in Thailand, the males, but never females, will have either two tusks or one, and sometimes apparently none. However, Thai tuskless males, called *si-dor*, actually have, in common with the females, tushes or short tusks, not visible beyond the lips. Tusks and/or tushes are present in miniature at birth, but only tusks continue to grow throughout the elephant's life at varying rates and into varying shapes (Fig. 40).

In the case of the Thai elephant, certain tusk shapes were traditionally considered desirable, their length and shape determining whether the elephant was trained for war or peace, whether it would transport a king or a commoner. To that end, traditional writings on elephants extensively classify and poetically describe the great variety of tusks, in terms of shape, thickness, length, direction of curvature, and even subtleties of colour. For example, a true *chang niem* or black elephant will have tusks the points of which are in the shape of banana blossoms (Boonsong Legakul and McNeely, 1977: 641). This has led to the development of a long-standing tradition of connoisseurship, and thus acquisition of auspiciously shaped, sized, and coloured tusks. Such prized tusks are usually of grand size and length, and suitably mounted on lotus-shaped ornamental pedestals, are displayed in pairs in wealthy Thai family homes, palaces, and offices of state. In the past, such tusks may have been acquired or inherited as relics of auspicious but deceased elephants. Unfortunately, nowadays tusks as symbols of wealth and status are still highly sought after, and it appears most likely that present-day elephants are being hunted for such trophies by the agents of unscrupulous *nouveaux riches*.

Ironically, while the freedom from domestication of the African elephant has led to its wide-scale slaughter for ivory, the relative docility of the domesticated Asian elephant makes it possible for long tusks to be periodically 'tipped' or sawn off safely, without damaging the nerves, allowing for further regrowth. Such tipping was traditionally carried out in the timber camps to lessen the tusker's potential danger to other elephants and men, as well as a means of increasing, when necessary, the effective use of the tusks in timber work (Fig. 41). This tipping also provided modest amounts of ivory for carving without resort to slaughtering the entire animal. Properly carried out, such tipping is not painful to the elephant.

In Thailand the one-tusked elephant (*ekathant*, one tooth) is very highly regarded (Fig. 42). A one-tusker is considered to have almost supernatural strength as it is a reminder of an elephant of

Fig. 41
Bath time in a forest camp. The stubby but well-developed tusks indicate recent tipping. (Photograph Monique Heitmann)

Fig. 42
One-tuskers or *ekathant* are highly regarded for their strength. (Photograph Rita Ringis)

legend, having the strength of a thousand, who with its solitary tusk slew a predatory ogress causing disharmony in the universe (Fine Arts Department, 1990). Further contributing to the respect in which a one-tusker is held is the fact that it resembles the one-tusked elephant-headed god Ganesha, the son of the god Siva. Hindu legend says that Ganesha tore out one of his tusks to use as a pen to write down the great Indian epic, the *Mahabharata*, as its composer declaimed it. Thus, in memory of that heroic deed, large or small votive images of Ganesha frequently portray him as holding that tusk and in one of his other four hands, the completed book of the epic (see Figs. 13 and 97). Episodes from that epic still feature in literature and art in Thailand.

Although the Asian tusker is assumed by the layman to be the exemplar in strength, and greatly valued by the timber industry for its ability to manoeuvre and balance massive logs on its tusks, in fact the *si-dor* or tuskless male frequently is the stronger and more agile, making effective use of its more highly developed trunk.

While the trunk of the African elephant is more wrinkled or horizontally ridged in appearance and terminates with two processes or finger-like appendages of great dexterity, the Asian elephant's trunk presents a less ridged appearance and has only one process or moist finger-like 'snout' at its tip (Fig. 43). However, this flexible process-tip makes it possible for the elephant to carry out delicate actions such as picking up berries or small objects. The sinuous undulating trunk of the elephant is a most versatile limb, combining incongruously the functions of a hand and nose. And what a useful limb it is, enabling the elephant to breathe and to call, to reach high and low, to suck in and to squirt out, to grasp and to hurl, to fondle and to caress. Holding its trunk aloft like an antenna, the elephant compensates for its poor sense of sight by picking up scents of impending danger or pleasure at great distances. While checking for future pleasure

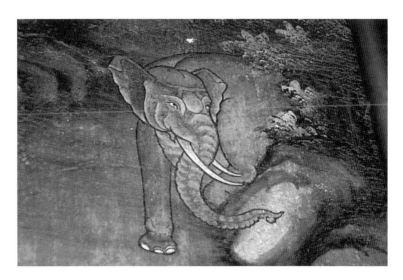

Fig. 43
Detail of a tusker, with the trunk terminating in a single process, characteristic of the Asian elephant. Mural at Wat Suthat, Bangkok. (Photograph Isabel Ringis)

might be acceptable, constantly sniffing the air apparently is not. Folk wisdom about elephants warns the potential buyer of the dangers to come with the purchase of a constantly 'cloud sniffing' beast (Toke Gale, 1974).

The trunk is also the elephant's most vulnerable and delicate appendage, something that it guards with particular care, tucking it carefully away when sleeping, as any damage to the trunk may endanger this great beast's life, for not only is the trunk essential in conveying food and water into the mouth, but also for breathing. Given this vulnerability, it is particularly significant that in play with family members, the elephant will constantly use the trunk to display what can only be perceived as affection and trust, placing its trunk into the mouths of other elephants, or playfully feeding little titbits to its companions. This obvious vulnerability has also given rise to apocryphal stories about the occasional elephant committing suicide by deliberately standing on its own trunk, thus cutting off the air supply.

It is also the trunk, as hand, that enables the elephant to be a tool-bearing animal, rare in nature. In fact, an ancient Sanskrit word for 'elephant' sometimes used in Thai writings is *hastin* which means 'an animal with a hand'. As a voracious but by no means indiscriminate eater, the elephant, using its trunk, will shake grasses and bamboos free of dust and mud before placing them into the mouth. Because of the pressing nature of the elephant's feeding routine (some 250 kilograms a day are necessary to satisfy hunger), it can simultaneously eat while gathering additional food with the trunk to place in its mouth when necessary. Working elephants in the forests are particularly adept at this.

To enhance its personal hygiene and sense of well-being, the elephant may use its trunk to grasp sticks or small bushes with which to scratch itself, or to slap at irritating insects. The trunk is also used to suck in and spray cooling water or soothing dust (elephant talcum powder) on to its sensitive hide. Clearly aware of the implications of the act, the domesticated Thai elephant is not averse to picking up and throwing dung at the veterinarian whose smell and appearance apparently remind the elephant of the regular vaccinations it must endure to maintain freedom from disease. Thai veterinarians interviewed during the course of this research particularly commented on pachydermal aversion to their presence.

On a grander scale in former times, the trained war elephant would use its trunk in battle to wreak havoc, dealing blows to those near it and dashing enemies to the ground (Plate 16). Apart from brute strength, conversely, the trunk of the elephant is capable of the most delicate tasks: gently coiling around its keeper, lifting him off the ground, and placing the keeper safely above, on the elephant's broad back.

The trunk's lithe dexterity causes people to perceive (not always correctly) the tamed elephant as playful, and to be drawn almost irresistibly, but not without a *frisson* of delicious fear, to make

Fig. 44
Timorous tourist feeding an elephant.
(Photograph John Ringis)

contact with that undulating appendage. In Thailand, the prevailing custom of offering delicacies such as bananas or sugar-cane to passing elephants (and they do pass, frequently) usually satisfies this mysterious human urge (Fig. 44). The elephant grasps the offering (and thankfully not the hand), and coils it upward into his mouth. But woe betide him who changes his mind about yielding the proffered delicacy. In extreme cases, this could, and has led to rage in the elephant, picking up the teaser, and dashing him to the ground. More usually, however, it results in a friendly tug of war, with the elephant always winning. The Thai proverb 'Ooy khaw paak chang' (The sugar-cane is already in the elephant's mouth) warns such a giver, both literally and metaphorically, that once something has been given away, it is impossible to retrieve it.

Until the early part of the twentieth century, the elephant was almost a daily companion or prominent presence in some form or another for most people in Thailand. As a result, the appreciative and fond perception of its qualities is deeply ingrained in the national consciousness, in language and in literature. This sense of affinity may be expressed in forms considered unlikely (and unexpected) in the West. For example, Thai admiration for the appearance of the trunk, and its supple and pliant motion, has led to poetic descriptions of it being incorporated into concepts of

45

Fig. 45
Walking bronze Buddha image, with pendant arm like 'the trunk of a young elephant', Sukhothai style, at Wat Benchamabophit, Bangkok. (Photograph Rita Ringis)

beauty and grace, both secular and sacred. Thus, when the supernatural physical appearance of the Lord Buddha is praised, his arms are described as resembling the trunks of young elephants. In turn, Buddha images, particularly those from Sukhothai of the fifteenth century, reflect this in their style, by clearly depicting those literary anatomical proportions (Fig. 45). Furthermore, in secular poetry, for example, the languid sway of a beautiful woman's walk may be likened to that of the rippling sinuous motion of the trunk of the elephant.

Contrasting with the muscular and graceful trunk is the elephant's restless and often stumpy and stringy tail, tipped by a brush of uneven bristles ... surely an apparent anticlimax. Yet, in the Thai notion, even the condition of the tail and the shape of its terminating hairs (whether a mere brush or handsome bodhi-leaf-shaped aureole) is but another indicator of the breeding and character of the elephant. The longer, straighter, and generally more comely the tail, the better the elephant. In fact, another proverb encapsulates folk wisdom about the desirable qualities of elephants' tails, and of all unexpected things, women: 'Du chang hai du hang, duu nang hai du mae, tha cha hai nae hai du yaay.' Freely translated, this suggests that when judging an elephant, you should check its tail; when choosing a bride, observe her mother, but if you want to be sure, take a good look at her grandmother!

Stringy tails aside, supporting the bulbous head and massive torso of immense weight are the elephant's relatively short pillar-like legs, with their small spongy pad-like feet protected by hoof or 'toe' nails, each of which may weigh up to some 5 kilograms. These nails, the condition of which may also help indicate the elephant's age and state of health, usually number eighteen, with four on each front foot and five on each back foot. (However, white elephants of the perfect kind must have twenty nails, five on each foot.)

These relatively stumpy legs look curiously inadequate, almost comic. However, they are again marvels of evolutionary balance and counterbalance. The front legs serve largely as supports, with the shorter stronger hind legs providing propulsive force when needed. Because of the great weight of the body, these hind legs, unlike those of other four-legged animals, cannot be tucked under the body when the animal lies down, as it must to sleep its necessary three to four hours in every twenty-four. Instead, the hind legs bend in a position similar to that of a human kneeling, otherwise the animal could not lift its great body weight from a rest position. This anatomical arrangement works very well with a healthy elephant. However, should a sick animal lie down, it is unlikely to be able to muster the strength to raise its vast body weight. Consequently, sick elephants, in their wisdom, fitfully doze and sleep standing up, propped against a tree or suitable support. One elephant, clearly unwell, met during the course of research for this book had not lain down to sleep for several weeks. Age-old observation of the balance of forces and vast strength necessary to achieve the manoeuvre of propelling some 3–4 tons of weight of the average Asian elephant up from a kneeling position and into motion is pithily expressed in a Thai folk saying that also illustrates traditional values, nowadays increasingly questioned: in a marriage or partnership, the woman is said to be like the hind legs of the elephant.

Overall, this pachydermatous or thick-skinned creature, contrary to that descriptive term, is very sensitive. In fact, it is a paradox that the elephant, shielded by its apparently thick hide further armoured by protective stubby coarse sparse bristles, is vulnerable to irritation in the extreme from minute insects taking up residence in the elephant's heavily grooved hide, impervious to the futile swishes of the stumpy tail. Regular bathing in rivers and wallowing in the mud helps relieve the itching caused by these mites. In fact, as in humans, the condition of the elephant's skin is an indicator of health and age. While skin the texture of a wrinkled custard apple may not be desirable in humans, such happy corrugations and furrows proclaim the health and beauty of a Thai elephant. Colour of the skin, as in humans so in elephants, also contributes to the general appearance of beauty (Fig. 46).

To the untutored eye, an elephant is essentially grey-coloured. In fact, variations of toning are evident throughout its life. A newborn's distinctive pinkish-grey-brown colouring is further

Fig. 46
Fair- and dark-skinned elephants from the Royal Elephant Stables. (Photograph Rita Ringis)

accentuated by an ephemeral dandelion floss of ginger-coloured hairs. The fully grown elephant's grey colouring may vary subtly in tone, with some elephants developing a pale sometimes spotty pigmentation around the outer edges of the ears and face (see Fig. 41). Subtle differences of colour in elephants of character are poetically enumerated in traditional writings, as well as depicted in painting (Plates 17 and 18). Such elephants are hailed as having the glow of molten silver, of burnt copper, of shimmer of running water, of rain-swollen clouds, of pink lotus, and so on (Fine Arts Department, 1990). White elephants, and other auspicious elephants, some of them 'black as the raven's wings', are discussed in detail in Chapter 5.

A curious feature of any elephant is the fact that its skin or hide appears to be too loose, a little too large as it were, for its body. This oversize skin may, to the anthropomorphic eye, increase the endearing and comic effect of the elephant, but it serves real purposes. For example, while the backside of an elephant is proverbially 'baggy'-looking, rather like a badly fitting diaper, it actually conceals and amply protects the scrotum. Such looseness may be particularly noticeable on the lower part of the legs. Around the 'wrist' or 'ankle' of the legs, these loose folds are clearly evident and function in a unique way. When the elephant puts its cushion-like footpad down flat, the loose skin and footpad allow the bones of the foot to spread, the better to distribute the weight. Conversely, as the elephant lifts the weight off its footpad, contraction of the foot-spread permits the elephant mobility in muddy ground where an otherwise inflexible foot supporting such great weight would bog it down. In a healthy elephant, this looseness or pliancy of skin is not excessive but subtle, and is thought to aid temperature control of the entire body. As well, it allows the elephant, by a sort of subcutaneous muscular shudder of its skin to shake off insects and other irritants from its extremely sensitive hide. The effect of this looseness of skin is particularly evident in the rippling and fluid grace of the elephant in motion, a visual paradox: vast bulk moving nimbly and almost buoyantly over varied terrain.

The elephant's agility in difficult terrain is proverbial, and led to its pre-eminent role in the transportation of people and goods in South-East Asia until early in the twentieth century (Fig. 47). Acute observations about its nimble-footedness under such conditions were made by Holt S. Hallett in the records of his 1876 journey, in the northern parts of Thailand, in search of a route for a possible railway to transport British goods from Burma to Thailand. Published in 1890, *A Thousand Miles on an Elephant in the Shan States* contains a wealth of fascinating detail about a now vanished world, while also documenting the author's victory over considerable discomforts endured.

At the outset of the journey, six elephants and their drivers were hired for the first stage, and forty small wicker baskets packed with stores for several weeks were stacked in the howdahs on the

Fig. 47
Transport elephants with covered
howdahs awaiting their passengers,
dignitaries on a visit to Khmer ruins.
From J. Thomson, *The Straits of
Malacca, Siam and Indo-China*, 1875.

elephants' backs. In addition to the stores in wicker baskets, also
carried on the elephants' backs were other essentials: 'The cooking
utensils, crockery, a dozen brandy for medicinal purposes, two
dozen of whisky, and some of the medicines ... two waterproof
bags and a tin box for clothes and money, my office box, rugs,
bedding, chairs, and camp bedstead, and our two selves....'
(Hallett, 1890: 5.)

Hallett's elephant's pitching and rolling gait, 'like a Dutch
lugger in a chopping sea', at first somewhat incommoded his
initial survey attempts, and he feared that his compass 'would
most likely jam ... into one's eye' (Hallett, 1890: 14). 'Yet by
giving way to the swaying movements of the brute' Hallett
achieved a satisfactory equilibrium. However, his admiration for
the elephants' skill was soon called forth:

The path, owing to the rain, was rendered slippery, and was so steep,
that the elephants at times had to slide down on their bellies, with their
legs stretching out behind and before them. To see these great clumsy-
looking brutes constantly kneeling down, crouching on their haunches,
and then rising again, as they ascended and descended the hill sides, in
order to keep their equilibrium and reduce the leverage; never making a
false step; putting one foot surely and firmly down before lifting another,
and moving them in no fixed rotation, but as if their hind and forequar-
ters belonged to two independent bipeds; every movement calculated with

nicety and judgement,—forced one to admire the sagacity and strength of the animals, and the wonderful manner in which their joints are adapted to their work. (Hallett, 1890: 25–6.)

Anyone who has ever ridden an elephant in jungle terrain, whether as part of an intrepid trek or even in the tourist shows of Thailand's elephant training camps, will agree that the experience could not be better expressed.

However, despite its relative agility, the elephant's mobility is somewhat hampered by the shortness of its legs in relation to its ponderous torso, making it impossible for an elephant to jump or leap even the shortest distance. In fact, it cannot take all feet off the ground at once. This apparently trivial deficiency in such a powerful creature makes it possible for man to keep it captive (or prevent it from entering cultivated areas) when necessary, by the simple expediency of digging a shallow ditch around the area in which the animal is to be confined, wide enough to prevent it from 'jumping' but narrow and deep enough to restrain it from plunging in to climb out again on the other side. Such techniques were sometimes employed during the hunting and corralling of elephants in the past (see Chapter 7).

A ditch may deter, but a stream or river will not suffice, for this vast seemingly land-bound creature is a great swimmer, paradoxically buoyant. This skill, coupled with the elephant's strength, was put to good use by the teak loggers in up-country Thailand. As a major means of transporting logs was by floating them downriver, elephants were employed to prevent log jams in the waterways and ensure that the teak logs were directed to their ultimate destinations.

In more recent times, the elephant's proficiency in swimming has been exhibited in a festive manner somewhat similar to the Surin Elephant Round-up, to draw tourists both from within Thailand and abroad, in elephant swimming races organized by the provincial authorities of Buriram. As in the Surin Round-up, the aims of the races are not only to generate profits for the province, but also to provide extra income for the almost destitute elephant keepers and their charges, whose daily search for fodder in areas now virtually denuded of their jungle coverage has become increasingly frustrating in recent years.

Of the Thai wild elephant, it can truly be said that it lives to eat and eats to live, spending its twenty waking hours out of the twenty-four largely searching for food, with pleasant interruptions for courtship and bathing. In those twenty hours, it may consume up to 250 kilograms of food, as well as 60 gallons of water. In the dry season, when streams may be low in water, the elephant ensures its daily intake by digging water-holes with its forefeet in sandy stream-beds, in effect sinking wells to the water-table, surely a sign of its much vaunted intelligence. As a herbivore, the elephant feeds on a variety of grasses, fruits, bamboos, barks, creepers, trees and shrubs, and unfortunately, given the opportunity, cultivated crops as well.

To fully appreciate the amount of food consumed on one given day in one given though extensive area, it should be remembered that wild elephants, except for the odd rogue, rarely move alone but stay in herds of up to twenty heads.[1] Domesticated working elephants consume the same amount of fodder, part of which is provided for them by their keepers, part of which they customarily search out themselves, being free to roam in the forests and jungles around their camps, though usually restrained from wandering too far by rattan hobbles on their front legs. Until the logging ban in 1990, such working elephants enjoyed the best of both worlds. In return for work in the morning hours, they were free to feed and rest in the afternoon in the cool and shady forests. This partial freedom has been long recognized as a necessity to maintain the health of an elephant. In fact, the closer a domesticated elephant is kept to its condition in the wild, the healthier it is likely to be.

A healthy elephant needs only about four hours of sleep out of the twenty-four and this it does, lying down on its side, usually in the cool night hours between about 11 p.m. and 3 a.m. During the heat of the day, it may occasionally briefly drowse standing up, propping itself against available trees or rocks. Sleep patterns and behaviour are also indicative of the state of health. A sick elephant will not lie down to sleep, but sleep standing. Such behaviour can be maintained for lengthy periods of weeks, and clearly indicates the elephant's awareness of its lack of well-being.

The health of domesticated elephants is usually closely monitored by their keepers who regularly provide them with salt-licks and various medications embedded inside the favourite treat of balls made of tamarind fruit paste. The known fact that wild elephants also crave and thus seek out natural salt-licks has led to the practice of placing these licks in strategic areas of national parks so that the coming and going of the wild elephants may be to some extent studied and monitored.

All mature male elephants, whether African or Asian, wild or tamed, are apparently subject to a mysterious not yet fully scientifically explained condition: the phenomenon of *musth*, which refers to a periodic state of riotously destructive and violent behaviour likened to madness. That this condition has not yet been fully documented in the wild in South-East Asian elephants is understandable, given the dangers posed by a rampaging bull elephant. However, some headway in its analysis has been made through observation and monitoring of the condition in tame working elephants. As soon as the condition, which is gradual, begins to manifest itself, such elephants are usually secured or chained.

---

[1]In the Thai language, quantifying nouns are used to denote numbers of particular things, objects, etc. For example, 'two books' is expressed by 'book two volumes'. In the case of elephants, the quantifying word is *cheuak* which signifies a rope, therefore 'twenty elephants' is expressed as 'elephant twenty ropes', a singular testimony to the antiquity of domestication of elephants in the land.

According to those who have had long experience of working with elephants, such as officials of the Thai Forest Industry Organization (FIO), this state occurs annually in the mature male elephant and lasts for about two to three weeks. Its onset is heralded by the appearance and gradual increase in volume of an oily pungent liquid secretion from a gland situated between the eye and ear on either side of the head of the elephant. Secretions in the genital areas also appear (Amnuay Corvanich, 1968).

In the fully-fledged condition of this state, elephants, if unrestrained, have been known to lay waste to their surroundings, and even to kill not only their mahouts (if these have been neglectful of traditional precautions), but also any human, animal, or even fellow elephant that gets in the way. Particularly dangerous is the case of several bull elephants in this condition at the same time in the same area. Interestingly, some historians have suggested that male elephants in *musth* were used to lead charges into battle in former times (Chula Chakrabongse, 1960). This condition of *musth* has been suggested by some to be comparable to the state of rut or heat in other animals. However, while there appears to be no doubt that the male in *musth* is sexually excited, and will mate, given the chance, it has been noted by those working with tamed elephants that the male also will mate when not in *musth*. Of interest are reports that cow elephants will avoid, if at all possible, a male in the full violence of *musth*.

Some authorities, including the FIO, have suggested that this violent state may be triggered by over-nutrition. Giving credence to this notion is their observation that this condition occurs more frequently during the hot season, when working elephants are on extended leave from their duties, and are free (though hobbled) to roam in the forest, and thus overindulge themselves with fodder. The FIO authorities also found that the onset of this condition may be partially controlled, and its duration shortened by special care which includes a regimen of hard work, and a diet which includes a particular type of green melon, thirty to forty a day (Amnuay Corvanich, 1976). After that dangerous period, most domesticated elephants will return to their mild docile state except for some who for whatever reason are habitually bad-tempered. This latter trait has been recorded in the traditional treatises on elephants, as well as observed scientifically.

Mating, as mentioned earlier, is not exclusive to the state of *musth*, and occurs spontaneously and apparently frequently at other times throughout the year. No doubt contributing to this is the fact that female elephants also exhibit a seasonal period of sexual readiness similar to *musth*. This condition is indicated by a similar secretion of fluid and a distinct odour which is apparently highly attractive to male elephants. However, the female in this state is not excitable or violent as is the male.

Given that many elephants are notorious for their reluctance to mate in captivity (a continual regret to many zoos), the love life of elephants is something of an elusive mystery to most people and perhaps best left unmentioned. However, it would seem that human curiosity about the sexual behaviour of elephants is reasonable, and not prurient, given the fact that elephants are known to have affectionate relationships with their companions and offspring, and as herd animals they have been observed to have extended relationships with other adult elephants. A graphic answer to such a reluctantly posed (but obviously universally tantalizing) question is illustrated in a generally available FIO pamphlet by a photograph with an endearingly candid caption labelled 'How Do They Do?' (Fig. 48).

Inspired imaginings in the distant past had fostered the idea that elephants mated facing each other, rather like humans, further increasing the elephants' legendary mystique and reputation for intelligence and wisdom, not to mention acrobatic skill. U Toke Gale (1974) notes that these misleading assumptions quite probably arose from the fact of the unique placing (for a quadruped) of the female elephant's reproductive organs at the same frontal position as that of the male's, and that the female elephant's breasts, unlike those of other quadrupeds, are situated on the chest, between the front legs. This is at the equivalent of the chest of a human. In effect, the female elephant, even though she is a quadruped, replicates certain features of the human female.

Any misapprehensions about apparently mysterious couplings of elephants are disabused by the clear and succinct descriptions given by Thai elephant veterinarians who point out that generally prior to elephant conjugal union, vigorous and affectionate fondling using the trunk may take place between the two animals. Subsequent to that foreplay, at a suitably appropriate moment, the male stands erect on his hind legs, maintaining balance by placing his forelegs on either side of the spine of his female companion, and after some flexible and lengthy adjustments consummates the process to mutual satisfaction.

Fig. 48
Surin Round-up elephants illustrate 'How Do They Do ?' (Photograph Rita Ringis)

53

Of relevance here is another myth long dear to humans about the matrimonial life of the elephant, concerning the notion of 'faithfulness' or monogamy in elephants. Perhaps contributing to this is the endearing and comforting stories that adults read to children of the uxorious Babar the Elephant and his sweet wife Celeste. Though it would be unwise to make generalized statements about the marital habits of Asian elephants in the wild, observations of the customs of working elephants would indicate that monogamy among them may be mythical. Though elephants are said never to forget, their partners if transferred to another camp will of necessity take on the next attractive comer. Serial monogamy would seem to be the mean, and that monogamy does not last long, given conception and the intervening forgetfulness during the eighteen to twenty-three months of gestation. Observations both in the wild and in tamed elephants suggest that once conception takes place, the cow elephant avoids male companionship, and the bull elephant is relegated to peripheral status. Towards the end of the gestation period, another cow elephant, or 'aunty', from the herd (or company) becomes a constant companion of the mother-to-be, and in the forest at the secluded birth, the aunty protects the vulnerable mother and newborn calf from predators (Amnuay Corvanich, 1976).

Generally speaking, the birth, whether of an Asian wild elephant, or even a Thai working elephant, takes place in the forest. The calf, weighing some 100 kilograms, is born enclosed in a thin transparent sac-like membrane (Amnuay Corvanich, 1976). This membrane contains substances that induce either the aunty or mother to free the calf by eating the sac, and thus also eliminating the evidence of the new birth, evidence that attracts predators in the wild. The aunty also supervises the calf's feeble and continual attempts to stand up. During this process, which may take up to a couple of hours, the mother is left alone to recuperate from her ordeal, before the calf is able to stand and suckle, directly with the mouth, not the trunk, at her two breasts. According to observations by the FIO, the period of recovery of the mother's equilibrium is essential, as instances of the mother killing a baby that stands 'too soon' have not been unknown (Amnuay Corvanich, 1976).

After the birth, the aunty may continue to assist in the rearing of the young, and should something untoward happen to its mother, it is likely that the aunty takes over the nurturing of the calf. During her lifetime, a wild cow elephant may drop up to four calves, one every three to four years of her fertile period, which begins at about sixteen years. However, this number is only an ideal in the case of many Thai domesticated elephants because elephant owners, either individual or company, were and are reluctant to lose the extensive working time of the mother elephant, taking into consideration the gestation period of up to twenty-three months, as well as suckling by the calf of up to three years (Cheun Srisavast, 1986). During that time, the mother elephant will be assigned to light duties, keeping her baby by her side, with both enjoying that brief carefree period (Fig. 49). After that duration, it is back to full-time work for the mother, and the young elephant calf must begin the lengthy training for the rigours of working in the forest.

While in the wild the herd cohesiveness is based on blood ties and relationships between the females and their young, in captivity these very ties must be superseded to make an elephant a trained and effective member of a working team. By tradition, the training was carried out in villages or forest camps by individual owners or owner-contractors who had amassed, through capture or trading, elephant 'herds' frequently unrelated by blood (Fig. 50). However, for many years a unique elephant training camp was situated in the spectacular limestone karst formation some 60 kilometres from Lampang, in northern Thailand. This was the Young Elephant Training School established in 1969 by the FIO. In its lush forest setting, the camp was also the retirement home or recreation centre for the many working elephants owned by the FIO. Many of the 'students' were the young of those elephants at the training school, which was regularly visited by interested observers and tourists to view the traditional elephant curriculum, which is outlined in Chapter 7 (Fig. 51).

Fig. 49
Cow elephant and calf. (Photograph Monique Heitmann)

Fig. 50
Elephants from all over Thailand performing in concert at the Surin Elephant Round-up. (Photograph Rita Ringis)

With the introduction of the logging ban in 1990, conditions for working elephants and their mahouts have changed dramatically and thus this institution has recently been 'replaced' as it were, or revitalized by the inauguration of the Thai Elephant Conservation Centre of Friends of the Elephant Foundation in 1992 to mark the Third Cycle Birthday Anniversary of HRH Princess Maha Chakri Sirindhorn, whose affectionate regard for the welfare of elephants is well known among the Thai people. Now set in spacious grounds near the Thung Kwien Community Forest some 35 kilometres from Lampang, it was established as a much needed documentation and information centre on the Thai elephant for researchers, as well as a refuge for the increasing numbers of domestic elephants in need of medical assistance and care. However, the Centre is open daily to tourists, to promote public awareness and understanding of the elephant's contribution to Thai life in the past and present.

While the working elephant's family ties are essentially subject to the whims of its master, the life pattern of the wild elephant is somewhat different. Wild herds consist mainly of females and their varied generations of young, and are led by old cow elephant matriarchs, with the previously sexually 'successful' bull elephants (in terms of fathering young) being relegated to the peripheries of the herds, as general guardians against predators. This female-dominated nurture pattern is evident in herds of elephants in the wild for both African and Asian elephants.

It is said that for both African and Asian elephants in the wild, as the calves mature, the young bulls gravitate towards the older male guardians and serve as their acolytes in generally protecting

56

Fig. 51
Mahouts training young elephants.
(Photograph courtesy of the Forest
Industry Organization)

from a distance the well-being of the continually young-nurturing herd, led by the venerable matriarchs. A hierarchy of these peripheral guardian bulls is also evident and appears to be a continually evolving process, through a younger bull challenging the dominant bull to combat. The winner then becomes the leader, while the loser drifts away further to the fringes of the herd, often becoming something of an outsider, and perhaps eventually becoming that most feared creature of the forest, the rogue elephant (Boonsong Legakul and McNeely, 1977: 637). Such a rogue may have been responsible for a particularly gruesome attack some years ago in one of Thailand's national parks on a photographer and his monk companion. The moment before death of both men is recorded in a photograph still in the camera, found afterwards at the scene.

Because much about the elephant's nature remains a mystery, anthropomorphic myths about elephants have flourished for centuries. An elephant never forgets, says the adage. This depends of course on what it should 'remember'. For example, not only does the elephant 'remember' what foods are succulent, but also which individuals caused it pain. At the Young Elephant Training School, the Thai veterinarian greatly devoted to his charges, Dr Pricha Puangkham, ruefully confessed that 'his' elephants clearly did not 'like' him, because they associated him with the dreaded vaccination needle which he wielded regularly for their continuing health. In the same way, during conversations concerning his reminiscences as a teak wallah in the 1950s, Dacre F. A. Raikes, a British-born long-term resident of Thailand, also noted that elephants in his camps at the time did not seem to like 'farangs' or foreigners, which the local folk attributed to their different or alien smell: in teak camps of the past, it was invariably the foreigners who wielded the needle.

These patterns of learned behaviour depend largely on simple conditioning, but suggesting actual complex reasoning in the elephant are numerous tall tales about gratitude expressed by elephants, or of long-nurtured and calculated revenge taken by them against their wrongdoers. Some of the earliest of these tales were recorded by French emissaries to the Court of Siam in the seventeenth century. One, perhaps recounted by a scoundrel attempting to account for his new-found and inexplicable wealth, tells of an elephant's 'gratitude' for having had a thorny spine removed from its foot. Apparently, the extraordinary elephant revealed to its saviour's delight a vast cavern filled with treasure which the elephant had collected by robbing passing travellers (Choisy, 1687).

In the same spirit, Simon de La Loubère noted that the Siamese 'do think that Elephants are capable of Justice', and recorded a story of one elephant whose dignity had been hurt by a man breaking a coconut on its head. According to the Siamese, whom La Loubère was loath to believe, the elephant bided its time, and eventually killed the offender, afterwards triumphantly placing a coconut on his body. Loath or not, it makes a lovely story (La Loubère, 1693: 46), as do the oft-told tales of domestic elephants filling with mud the wooden clappers that customarily hang around their necks, so that no noise is made as they steal into village plantations for midnight feasts.

Scientists are always understandably reluctant to make 'connections' that have insufficient objective data to back them up, particularly in the case of attribution of human emotions, such as sorrow or remorse, to elephants. For example, one veterinarian during the course of this research mentioned the case of an elephant thought by many to be clearly mourning and 'weeping' after killing his keeper. Even so, as a scientist, he confined himself merely to the observation of the contiguity of the events. However, conditioning and contiguity aside, here should be mentioned the frequently observed mysterious behaviour of wild African elephants in fondling the bones and tusks of their dead relations (Moss, 1989). This behaviour has not been so far documented in the Asian elephant.

Contrary to the myths about an elephant's lifespan reaching up to 150–200 years, the length and stages of life from its babyhood, childhood, adolescence, prime, and decline in terms of years closely parallel those of humans. Like the human, the elephant lifespan may reach the biblical three score years and ten or thereabouts, with the prime, in terms of the working elephant, being reached at about twenty-five years of age and lasting until about the age of fifty, beyond which decline sets in. After that time, working elephants are given gradually less strenuous tasks, and are retired from active work at about sixty, being then free to enjoy their remaining time roaming about with their mahout in attendance. On the death of a working elephant, cremation or burial is usual, the latter more likely if the elephant dies naturally of age, the former if it dies after an illness.

However, elephants in the wild, never having been subjected to the lifetime of strenuous work of a timber elephant, may enjoy a longer lifespan. In the case of the death of elephants in the wild, myths again abound, as men have claimed to have found 'elephant graveyards' to which the dying elephant migrates, alone, as though answering some mysterious call beyond the scope of present understanding. Some writers on elephants suggest that these so-called graveyards are mere bones that have collected at blockage points on river paths, bones no doubt of elephants, but bones of elephants that have died alone, near river areas where they may have gone when sick, and become trapped by their weakened state in the mud. After death, their bones have been swept up by floods and collected along the length of the rivers, to be brought together in vast numbers, at natural catchment areas, giving the impression of a graveyard of elephants' bones.

No matter whether myth or logic is correct, the individual death of such a vast and powerful creature has an awesome dignity. An awareness of the magnitude of such an event is inherent in the Thai language in that the common word for dying, *taay*, is not used in the case of an elephant's death. Instead, an elegant (and nowadays increasingly archaic) word, *lom*, is used, meaning 'to topple or keel from an upright position', and in context, signifying the death of an elephant (Haas, 1976). However, even more awesome is the currently threatened possible extinction of the entire species.

4　Elephants in Thai History

LONG known in the West as the Land of the White Elephant, Thailand has a complex cultural heritage and history in which, not surprisingly, elephants have played a conspicuous role. The documented history of any nation is a record of the dramatic interplay of human affairs, complex events the significance of which may be variously interpreted, depending on the historian's bias. While comprehensive interpretations of Thai history are to be found in numerous scholarly works, some of which are listed in the Bibliography, the thumb-nail sketch that follows below does not pretend to approach an exhaustive examination of the intricacies of that history. Rather, it selectively highlights some of the threads in the major themes, focusing on pivotal and legendary moments, enlivened, where possible, by contemporary or near-contemporary accounts of various events, as they relate to the subject of elephants in Thailand. Many of the observations that follow are based on a variety of source materials, few of which are definitive in terms of elephants at least.

Since elephants were a major component in traditional battles in South-East Asia, this history will seem biased towards the warlike and the martial. This, however, is not to say that the arts of peace did not flourish in the development of Thailand. Indeed, it is in fact through examination of the numerous monuments of the arts of peace, those of temple architecture and sculpture, as well as those of decorative embellishment, including traditional woven silks and cottons, that we may glimpse the substance of now vanished societies, their customs, and not the least, the role played by elephants in peace as well as war (Plates 19 and 20).

Possibly the earliest visual depiction of Thai people dates from the early twelfth century and is a crisply carved stone bas-relief at Angkor Wat, in Cambodia (Fig. 52). Identified by an accompanying inscription as 'Syam ...', they are depicted, with their Khmer counterparts, as part of a dynamic though disparate battle formation marching through a luxuriant forest (Coedès, 1968: 190). Towering above both groups of these foot-soldiers is a caparisoned war elephant, kept in check by an officer, holding a long elephant goad over his shoulder. Poised to his rear, on the elephant's back is an archer. Both officers, shaded by umbrellas denoting their noble rank, are clad in outfits similar to those of their Thai ('Syam') foot-soldiers below.

60

Fig. 52
'Syam' and Khmer warriors. Detail of a bas-relief at Angkor Wat, twelfth century. (Photograph Eileen Deeley)

The composition of the scene is at once masterly and amusing, for the demeanour of the foot-soldiers provides a pointed comment, albeit ultimately ironic. The Khmer soldiers, distinguishable by their austere uniforms, are disciplined, marching in rigid military formation, while their 'Syam' or Thai counterparts, vassals at the time and obviously considered 'barbarians' to a man, are not only apparently gossiping but also totally out of step and out of order, their head-dresses and lances askew. Perhaps this amusing observation illustrated not barbarism but an independence of spirit, presaging the eventual successful revolt of the Thai against their Khmer overlords sometime in the mid-thirteenth century, at the then Khmer provincial outpost of Sukhothai in the upper Central Plains of present-day Thailand.

With Thai independence asserted, a new era began at Sukhothai, which name is aptly translated as 'the dawn of happiness'. Much of what we know today of this early period of Thai history is inscribed on the four faces of a stone obelisk, believed to be the earliest document using Thai script, and traditionally accepted as having been composed in 1292. The inscription on the obelisk or stele, found many centuries later in the ruins of Sukhothai, describes life in the newly burgeoning kingdom.[1] The archetypal quality of the tale told on the stone, vivid in its simplicity and clarity, has led to its becoming an enduring vision of a golden age, and has a deep hold on the Thai imagination, rather

[1]The stone obelisk (today known as the Ramkhamhaeng Stele or Inscription No. 1) was discovered in Sukhothai in 1833 by the then monk-scholar-prince Mongkut, who was to become Rama IV in 1851, the king known to the Western world as the king in the make-believe musical 'The King and I'. In recent times, a group of foreign and Thai scholars have questioned the authenticity of the 1292 date and the contents of the inscription in what has become a major and continuing controversy in the academic community.

like the vision of Camelot, another golden age in another world.

Every Thai, young or old, is familiar with the poetic descriptions on the stele, of this former age of abundance and peace under a just king:

This land of Sukhodai is thriving. There is fish in the water and rice in the fields. The lord of the realm does not levy toll on his subjects for travelling the roads; they lead their cattle to trade or ride their horses to sell; whoever wants to trade in elephants does so.... (Griswold and Prasert na Nagara, 1971: 205–6.)

The inscription further reveals that the possession of elephants was of considerable importance in the economy of this early Thai city-state. Acquired by hunting (by lasso or being driven into an elephant corral), or captured as booty of war in raids, elephants were evidently regarded as sources of wealth and symbols of status. In fact, regional or vassal rulers associated with the new kingdom are defined by their riding and owning elephants, and those dispossessed of their fiefdoms as having no elephants. Corroborating the value accorded to elephant ownership is a telling order of precedence recording the laws of inheritance—when a commoner or man of rank dies, his estate, including his elephants, wives, children, granaries, groves, and so on, *in that order* passes in its entirety to his son (Griswold and Prasert na Nagara, 1971: 206–8).

The inscription also documents that in Sukhothai and on its outskirts monasteries and temples to the Buddha abound, and that ceremonies of homage to the Buddha are regularly conducted by the people and their king, mounted on his caparisoned white elephant whose tusks are decorated in gold. The ruler of this kingdom, as described on the stele, is regarded by the Thai people as the prototype of the just and benevolent king. His name, Ramkhamhaeng, means Rama the Bold, a name he earned, according to the inscription, as a young man in a mighty elephant duel defending his father's kingdom against a powerful enemy, when all others had fled the scene in disarray. Throughout much of subsequent Thai history, such combat by leaders mounted on war elephants appears to have become a customary and dramatic way of deciding dynastic and territorial disputes (see Plate 13).

Another famous and decisive duel was fought nearly 200 years later, in 1424. By then, the brief glory of Sukhothai had waned, and the city had been relegated to vassal status and provincial decay. Power now resided in the Central Plains, in the rising might of the new Siamese kingdom of Ayutthaya, which had been established in 1350. The duel was for no less than the throne of Ayutthaya, between two sons of the recently dead king. Such extremes of sibling rivalry are not uncommon in Thai history, as the kings from earliest times practised polygamy, and it was only in the early twentieth century that the concept of primogeniture fully came into customary consideration for inheritance of the throne.

Mounted on tusker elephants and pitted against each other for the throne of Ayutthaya were two brother-princes, aptly named Prince One and Prince Two. As both were mortally wounded in the duel, their brother, Prince Three (Chao Sam Phraya), became king.

To commemorate the battle, and also the cremation site of his brothers, the new king built a temple, Wat Rajburana, which is today one of the most famous and visited tourist spots in Thailand (Fig. 53). Sealed within the crypt of the *prang* or sanctuary tower of this mid-fifteenth-century temple was a veritable treasure trove of golden ceremonial objects, including a bejewelled kneeling elephant (see Plate 19). These golden treasures, unearthed in the 1950s by the Fine Arts Department, are now on display at the Chao Sam Phraya National Museum in Ayutthaya.

Ayutthaya was destined to develop into a great commercial entrepôt in the South-East Asian region over the next three centuries. Yet during that time, placed as it was between warring Burmese states on its west, and the once-mighty Khmer empire on its east, Ayutthaya was also involved in almost continuous wars, defending its northern tributaries and central regions from repeated incursions, or asserting its sovereignty time and again over regions then deemed vassal, present-day Laos and Cambodia.

Fig. 53
Wat Rajburana, Ayutthaya, built in 1424 on the cremation site of two princes killed duelling on elephant-back. Engraving from Henri Mouhot, *Travels in the Central Parts of Indo-China . . .*, Vol. 1, 1864.

The results of these conflicts were rarely decisive in the long term, though benefits were to be gained in the capture as slaves of the many unfortunate peoples and their possessions in the way of the advancing or retreating armies. For example, periodic stands against the Khmer at Angkor, while contributing to the sad decline of that civilization, brought benefits to the Siamese, not the least being captured Khmer war elephants and their expert Khmer trainers.

The inconclusiveness of these continuing and drawn-out states of war was partially due to the nature of political organization of the entire region, with its individual city-states forming dynastic and strategic alliances (usually short-lived) to extend their geographical power base, thus challenging the central power of Ayutthaya. Such challenges were usually defeated, either in battle or paradoxically by geography and climate. Ayutthaya was set at the confluence of three major rivers, and during the annual flooding of the Central Plains, the city became essentially an island fortress, into which the surrounding rural food-producing population could retreat, leaving any siege army to flounder and grow weary in the mud, to retreat, or to starve.

However, in actual battle, the strength of the Siamese army lay in the deployment of elephants and men in various special configurations traditionally considered effective, depending on the lie of the land and whether the battle was defensive or offensive. These configurations followed principles outlined in *Tamra Pichai Songkram* [The Treatise on War Strategy], compiled in 1518 (Rong Syamananda, 1977: 41). Of this original document only relatively recent late eighteenth-century illustrated manuscript versions revised in the nineteenth century exist today (Quaritch Wales, 1952: 121). Each distinct formation was in effect an ideal battle plan, and was identified by the name of a mythical and auspicious creature, for example, the lion configuration was called Singhanam; the Garuda configuration, Khrutnam (Fig. 54). This latter formation, for example, consisted of successive lines of infantry, foot-guards, and lance-bearers, followed by cavalry, beyond which was the commander-in-chief (mounted on a war elephant) at the centre of the formation, flanked by other war elephants. Protecting the rear were further groups of cavalry and infantry. Also taken into consideration before joining battle were numerous astrological predictions (such as auspicious cloud formations auguring well) and various sympathetic magic practices (Quaritch Wales, 1952; Saeng-Arun Kanokpongchai, 1988).

The scholar H. G. Quaritch Wales, in his *Ancient South-East Asian Warfare* (1952), outlines the various animistic, Buddhist, and Hindu-based propitiatory ceremonies performed before armies joined what he termed 'idealistic' warfare. Such warfare was dominated by ceremonial rather than more pragmatic practices adopted increasingly after the seventeenth century, following the introduction and widespread use of gunpowder. However, prior to that time, the utmost importance of the elephant and the

Fig. 54
Battle formations outlined in a *khoi* paper manuscript of 'The Treatise on Military Strategy', nineteenth century. (Photograph courtesy of the National Museum, Bangkok)

skill of its rider in battle was clear. Quaritch Wales refers to the Siamese Palatine Law, dating from the mid-fifteenth century, which outlined various rewards and punishments in battle. One who defeated his adversary while on elephant-back could be rewarded with golden garments and a high rank, while one who killed his opponent or his elephant would receive not only golden vessels and garments of rank, but also a wife. Ordinary foot-soldiers helping their mounted master would also receive many benefits. On the other hand, anyone going into battle and falling the length of three elephants behind his expected position would be executed. To discourage anyone fleeing from battle on his elephant, the Palatine Law stipulated that not only this coward but his entire family should be destroyed. In battles on the water, war barges and their oarsmen falling too many boats' length behind in battle formation would find themselves in chains or executed, or perhaps sent for life to cut grass for the numerous and hungry war elephants of the realm (Quaritch Wales, 1952: 186).

Of later battle formations, a seventeenth-century foreign observer noted that traditionally some nine battalions were deployed, each of which had sixteen male war elephants. Each of these elephants carried a standard, and was accompanied by two female elephants, to give 'dignity to the males' but also making them easier to control (La Loubère, 1693: 91). In these forma-tions, present as well were scores of transport elephants, bearing supplies. However, unlike the transport elephants which could be controlled by one man seated at the neck and applying directional pressure behind the ears, a war elephant would be ridden by at least two men, one at the neck and the other positioned towards the rear of the elephant, above the crupper, as can be seen illus-trated in some of the traditional manuscripts and mural paintings depicting battle scenes. Two men were essential, given the neces-sity of quickly manoeuvring such vast creatures, depending on vis-ibility both in front and behind. It should be noted here that

65

unlike a horse, an elephant cannot be bridled. At work, whether in battle or ceremonial, the chains or ropes that form part of its accoutrement (passing across the chest, under the belly and linked to the crupper along the lower back) are essential only to stabilize whatever is to be secured on its back, be it a howdah or mere baggage. For immediate control of the animal, each man would be armed with a lance-like spear that has a sharp hook-like attachment which is used to goad the elephant at various pressure points of its body. When firearms came into use, a third man, armed with a weapon, would be seated on the elephant's back, between the two 'controllers' (see Plate 12). The firearms no doubt contributed to the unpredictability of the battle, given the sensitivity of elephants to sudden noise.

This arrangement of riders, however, was different when it came to mounted leaders of battalions, divisions, or even whole armies. In that case, at least three, sometimes four men were carried on each war elephant. Seated at the neck of the elephant, above the fray of battle, would be the commander, sometimes even the king. From his position of vantage, he could make quick decisions as to the continuing deployment of troops. He would convey his decision to a signaller, who was seated directly behind and above him on a howdah, and thus clearly visible to the troops behind and below (Plate 21; Fig. 55). In fact, because of its height, the elephant was essential to communications. The signaller would convey the leader's orders to other strategically placed commanders by signalling appropriately with an array of peacock feathers held in each hand, raised high above his head (see Fig. 62). Also carried on the general's howdah were various handsome though clearly lethal weapons, ready for use, set out rather like an array of surgeon's intruments. Seated at the back of the commander's elephant, near the tail, would be either one or two mahouts or controllers, armed with goads to assist the leader in manoeuvring the beast.

In fact, the tusked war elephant, often partially covered by a protective shield of treated buffalo hide, was an early form of the armoured tank in battle. Such 'tanks' were particularly useful in sieges of walled towns or enemy stockades, most of which had tall 'elephant gates', to allow their own elephants entry and exit (Fig. 56). Ironically these were the most vulnerable points of a walled town or stockade, given the butting power of an armoured elephant. However, for that reason, defensive measures were implemented, and even today, at the Grand Palace area in Bangkok the main gate (popularly known as Pratu Chang: the elephant door) has sharp metal studs on its panels (Fig. 57).

The best-laid plans of battle leaders could be thrown into total confusion by various tricks designed to craze the elephants and set them rampaging amongst their own troops. A favoured and simple method of creating confusion and chaos was the use of spiked ball projectiles, thrown on to the ground for the elephant to step on. As the elephant's feet are very sensitive, the results of

Fig. 55
Royal war elephant re-enactment, with commander-signaller seated on the howdah, secured to which are various weapons. Soldiers with shields protect the elephant's legs. (Photograph courtesy of the Tourism Authority of Thailand)

Fig. 56
Elephant gate entrance to Wat Phra Ram, Ayutthaya. (Photograph courtesy of the Tourism Authority of Thailand)

such tricks in the heat of battle, with thousands of men and elephants, can be imagined. To protect the animal from such enemy interference from below, each leg of the elephant would be flanked by vigilant men. Their vigilance was ensured not only by their being attached by loose bindings to each leg of the elephant but also by the threat of the death penalty for any dereliction of duty.

Despite formidable armies of defence, the mid-sixteenth century was a time of turmoil for the people of Ayutthaya, weakened by a series of successful Burmese incursions in the northern regions, and internal dynastic power struggles at home. These came to a dramatic and byzantine climax in 1548, when the throne was seized from its boy-king incumbent by the regent, lover of the boy's mother, a concubine of the late king, who himself had been a usurper, though acceptable to the people, as he had been of noble birth. Reacting quickly, a group of princely conspirators

Fig. 57
European- and Thai-style elephant gates
at the Grand Palace, Bangkok.
(Photograph courtesy of the Tourism
Authority of Thailand)

spread a (false) rumour that a white elephant had been seen in the forest, near the capital. White elephants were considered to signify divine approval of a monarch. Thus, eager to secure it, for this would augur well and imply legitimacy of his reign, the usurper-king and the lady, by now his queen, set out to search for the animal, and were promptly ambushed and dispatched by the conspirators (Rong Syamananda, 1977: 47). In all, there were to be thirty-four kings of Ayutthaya, several of them in fact usurpers, but only thirty-three are officially acknowledged today, and this regent-turned-king and his perfidious queen live on only in infamy.

However, by a strange stroke of fate, it was not long before a historic heroic act redeemed the reputation of royal ladies. In 1549, the Burmese, hoping to take advantage of the still troubled Ayutthaya kingdom, invaded Siam through the Three Pagodas Pass and marched to the very gates of Ayutthaya. The new and legitimate ruler, King Maha Chakraphat, his two sons, and a retinue which included his queen and daughter disguised as noble warriors, all on elephant-back, emerged from the fortified city, to assess the enemy's strength. To minimize the slaughter of a general battle, the king accepted a challenge from a Burmese general to fight a duel on elephant-back.

During the chaos of the battle, as the two gigantic tuskers feinted and charged at each other, the king was placed at a dangerous disadvantage. Quick as a flash, two gallant warriors on elephants broke from the ranks, interposing themselves between the king and his assailant. The fatal blow meant for the king was sustained by one of the warriors—none other than the queen, Sri Suriyothai, whose daughter was also mortally wounded in the fray. Queen Sri Suriyothai's selfless bravery has become legendary, and her courage has been memorialized by a handsome *chedi* or Buddhist monument in Ayutthaya. The battle scene was popularized in late nineteenth-century Thai epic painting. Prior to that time, traditional painting was mainly confined to religious subjects and themes.

Despite their tragedy, the Siamese at Ayutthaya held out for some months. Though the Burmese soon withdrew, accepting two war elephants as reparations, this proved to be an unfortunate precedent. Thai legend has it that these two elephants, loyal only to their former masters, caused many headaches for their new Burmese owners.

In the ensuing years, King Maha Chakraphat devoted himself to strengthening his defensive capabilities and fortifying the capital and provincial outposts against future attacks from Burma or sporadic raids from Cambodia. Major elephant hunts were held, lasting many months in the forests and capturing hundreds of elephants for future warfare. Of all these captured elephants, seven were white, a large and auspicious number (Rong Syamananda, 1977: 49).

Naturally, these augured well and enhanced the king's power in the eyes of his people. For this reason, King Maha Chakraphat was acknowledged by all as 'The Lord of the White Elephants', and envy of this title was in part responsible for the Burmese attacking Ayutthaya in 1563. Prior to that, the new king of Pegu in Burma, King Bayinnaung, known popularly as Hansawadi, had made an apparently simple but in reality outrageous request: that the crown of Ayutthaya should present him with two of the seven white elephants. This request was no doubt prompted by the Burmese king's desire for such symbols of legitimacy, to enhance his own claims to be recognized as a Chakravartin, or World Emperor, the prototype of the just Buddhist ruler, who is the possessor of such auspicious elephants. This title was held sacred by both Burmese and Thais. However, only one ruler in any given era could be considered to be a Chakravartin, hence the conflict of interest (see Chapter 5).

The Siamese king's advisers were adamant in their opposition to this request. One legend has it that King Maha Chakraphat of Ayutthaya, to mitigate the bluntness of the refusal, pointed out to his Burmese brother-monarch that a great ruler has at his court various appurtenances of majesty, signifying his might and righteousness. These included, naturally, white elephants as well as bevies of fair ladies (who were in fact 'hostages' to their families' loyalty and vassalage to the sovereign ruler). Thus it was duly and most courteously suggested to Hansawadi, the ruler of Pegu, that surely he also, as a righteous monarch, had such possessions of privilege at his court, but that an exchange of white elephants between the two kingdoms would be as inappropriate as an exchange of beautiful ladies (Fine Arts Department, 1990). This was not received well by Hansawadi, and in 1563 he invaded Siam, and laid a long and bitter siege to the city of Ayutthaya. To alleviate the city's suffering and release it from the onslaught, King Maha Chakraphat sued for peace and was forced to hand over no less than four white elephants as well as nobly born hostages. One of the hostages was a young nine-year-old prince, Naresuan, son of the governor of Phitsanulok, a strategically crucial outpost of Ayutthaya. Within five years, despite the fact of such appeasement, the city of Ayutthaya, weakened by continuing internal divisions, disloyalties, and betrayals, fell to a successful Burmese siege, in 1569, ending the dynasties that had ruled since 1350. With Ayutthaya now a tributary of Burma, the Burmese installed a new vassal king, none other than the father of the young hostage prince, Naresuan, who had grown up in Burma (Fig. 58). Naresuan was now fifteen years old and was allowed to return to Ayutthaya.

During his exile in Burma, the young Naresuan, though a hostage, had grown and played side by side with the Burmese Crown Prince. Together they had learnt the noble arts of the princely warrior, and the Siamese prince had become thoroughly familiar with Burmese ways of thinking, perhaps misleading the Burmese into assuming that he was sympathetic to their ways. However, on his return to Ayutthaya, Naresuan was proclaimed Uparaja or deputy king, and he immediately began methodically and vigorously to prepare himself and his people for the restoration of the independence of Siam, which he declared unilaterally in 1584. Naresuan's diplomacy in winning over the allegiance of former allies of the Burmese, coupled with his military tactics of using guerrilla-like warfare and a scorched earth policy ensured that the Burmese, though vastly outnumbering the Siamese army, were continually harassed and handicapped in their efforts to subjugate Siam again. In fact, Naresuan is considered to have modified (for the first time) the 'idealistic' traditions laid down earlier in *Tamra Pichai Songkram* (Quaritch Wales, 1952: 121).

Naresuan's personal courage and unorthodox methods, for the time, became legendary: in leading an assault against a Burmese fortified camp, to inspire his perhaps understandably reluctant men, he scaled the stockade ahead of them all, with his sabre clenched between his teeth (Fig. 59). However, such feats, though undeniably brave, did not lead to conclusive results. In 1593, in response to continuing Burmese incursions, Naresuan (by now king) and his army prepared to rout them in the west, near the town of Suphanburi. In the heat and dust of battle, King

Fig. 58
The young Prince Naresuan (*right*) with the Burmese Crown Prince. Mural at Wat Suwandararam, Ayutthaya, nineteenth century. (Photograph John Ringis)

Fig. 59
Prince Naresuan, sword clenched in his teeth, scaling the walls of an enemy fort. Mural at Wat Suwandararam, Ayutthaya, nineteenth century. (Photograph John Ringis)

Naresuan, mounted on his war elephant, found himself unprotected by his usual infantry guards. His war elephant, which is said to have been in *musth* at the time, and thus more ferocious than usual, had charged in a rage into the midst of the enemy's ranks, bringing King Naresuan face to face with the enemy's leader, the Crown Prince of Burma, his former 'brother' and classmate during his childhood as hostage in Burma. King Naresuan hailed his 'elder brother' and challenged him to hand-to-hand combat on elephant-back. Being in the decidedly superior position, the Crown Prince of Burma could have refused and ordered his troops to kill his enemy, 'brother', and former friend. However, responding chivalrously, he accepted the challenge. This elephant battle is a highlight of Thai history, and there are many dramatic versions of it still recounted today (Fig. 60). Some say that King Naresuan's elephant was smaller than that of the Crown

Fig. 60
Elephant duel between King Naresuan
and the Burmese Crown Prince.
Memorial monument at Don Chedi,
Suphanburi, twentieth century.
(Photograph courtesy of the Tourism
Authority of Thailand)

Prince, and that Naresuan had to exhort it to courage and glory by chanting powerful odes into its ear and pouring magic unguents on its trunk to encourage it to charge (Van Vliet, 1640: 81). Be that as it may, when the two massive tuskers joined battle, the Burmese Crown Prince slashed at King Naresuan with his lance, barely missing the king's head but slicing off a crescent from the brim of his leather hat. In the next charge, as Naresuan's elephant plunged its tusks into the underbelly of the opposing elephant, the Crown Prince was caught off guard, and mortally wounded. The slain prince was carried off the field, and shortly afterwards, the Burmese retired from the fray (Chula Chakrabongse, 1960: 49).

To commemorate this victory, King Naresuan is said to have had a monumental *chedi* erected just outside of Ayutthaya at Wat Yai Chaimongkol (Temple of the Great Victory). Even today, 400 years later, this towering structure (in its restored state) is clearly visible on the horizon for miles before the traveller or pilgrim reaches Ayutthaya. Furthermore, on the actual site of the battle at Suphanburi, King Naresuan had a smaller *chedi* built, commonly known as Yuthahatthee. This was unusual in that it was not constructed, as is customary for a *chedi*, as part of a temple, but as a place of pilgrimage commemorating the elephant duel, and perhaps also the king's own subsequent act of forgiveness (Fig. 61). In fact, this

72

Fig. 61
King Naresuan pouring lustral water, signifying meritorious action. Memorial monument at Don Chedi, Suphanburi, twentieth century. (Photograph courtesy of the Tourism Authority of Thailand)

monument, according to the twentieth-century father of Thai Art History, HRH Prince Damrong Rajanubhab, was built at the suggestion of the Supreme Patriarch of the Buddhist clergy of the time who had appealed to King Naresuan to acquire merit by sparing the lives of the twenty-five men who, according to military law, were to be executed for failing to protect the king and his war elephant, and thus by their negligence, causing the king to come, unprotected, face to face with the Burmese Crown Prince (Apinpen, 1988: 91).

King Naresuan's victory and reign are still vividly remembered and honoured by the Thai people in various ways. In recent years, at the site of the original Yuthahatthee *chedi*, now restored as Don Chedi in Suphanburi province, a great annual fair is held, beneath a monumental sculpture of King Naresuan (Fig. 62). This modern work in bronze depicts King Naresuan seated at the neck of his massive war elephant, with his triumphant signaller seated on the howdah behind. This same scene is depicted on the back of the current 100-baht Thai banknote, accompanied by the words, in Thai script, 'King Naresuan the Great, BE 2098–2148' (AD 1555–1605).

To aid him in his struggle, King Naresuan had carried with him, as was and still is customary for soldiers, many tiny Buddha images. These were set around the crown of his hat. Today those

73

Fig. 62
King Naresuan, seated at the neck of his
war elephant, with a signaller seated on
the howdah. Memorial monument at
Don Chedi, Suphanburi, twentieth
century. (Photograph courtesy of the
Tourism Authority of Thailand)

treasured images are part of the Grand Palace Collection. In fact,
even the famous hat with the crescent cut in its brim, along with
King Naresuan's sword, was used as part of the royal regalia until
1767, when Ayutthaya was destroyed by the Burmese. However,
the same styles of articles of regalia, particularly the distinctive
hat, are still familiar to the Thai people from a portrait of the
present king of Thailand, His Majesty King Bhumiphol Adulyadej,
depicted with this regalia. Copies of this portrait are seen all over
the land, as it is a Thai custom for private households, offices, and
even shops to prominently display various portraits of royalty.

King Naresuan's epic victory in the elephant duel is also commemorated each year on 25 January, as Armed Forces Day. In his
brief reign of fifteen years, King Naresuan had restored Siam's
independence, and built up its military prowess, so that its might
was feared and respected as far as Laos in the north, Cambodia in
the east, and Burma in the west. Notably, the Burmese ports of
Mergui, Tavoy, and Tenasserim were made subject to Siam, facil-
itating access to the kingdom from the Indian Ocean, thereby
further opening the country to trade with the West.

Already from the early sixteenth century onwards, Ayutthaya
had seen the gradual advent of the Europeans, seeking new territ-
ories, trade, spices, and converts to Christianity. Trade treaties
with the first Europeans there, the Portuguese, had been concluded
to mutual satisfaction. The Portuguese, like subsequent European
arrivals, were given not only rights to reside and trade in the

74

region, but also to practise their religion. In return, they were to provide arms and ammunition for the defence of the realm. Unfortunately, the Portuguese provided these for both Siam and Burma, then traditional enemies (Rong Syamananda, 1977: 45).

Throughout the seventeenth century, Ayutthaya consolidated itself as a great commercial centre, to which traders and adventurers flocked from all over the world. The Portuguese had been followed by the Spaniards, the Dutch, and the English. Even the Danes from the far north were present, trading through the Danish East India Company firearms to Siam in exchange for elephants, some 300 a year, to be exported to India where they were used for war (Rong Syamananda, 1977: 67).

Not only European daredevil traders gathered in Ayutthaya. Arabic, Moorish, Persian, and Indian communities also contributed to the vitality of this trading city (Fig. 63). So did the Chinese, long present in considerable numbers and trading extensively though more unobtrusively than the Europeans. China was the only country to which Siam sent tribute, but this was merely a formality of long historical standing, as the Chinese were not interested in interfering in Siam's internal affairs, as were, increasingly, the Europeans. It would be reasonable to assume that some of the ivory, which at that time formed part of every noble Chinese lady's toilette and adornments, came from the quiet but elephant-rich 'vassal' of Siam. Ironically, these diverse foreigners resident in Ayutthaya until the mid-eighteenth century are immortalized (often comically) in mural paintings of a later date as part of the demon army in traditional depictions of the enlightenment of the Buddha (Fig. 64).

Writing in 1636, the Dutch East India Company agent in Ayutthaya, Joost Schouten, provides a vivid description of life in the kingdom at that time, listing the main provincial towns, and commenting that 'this royal and admirable City is perfectly well seated, and populous to a wonder, being frequented by all nations'. He notes that the city has over 'three hundred fair Temples and Cloysters, all curiously builded, and adorned with many gilded Towers, Pyramids, and Pictures without number' (Caron and Schouten, 1671: 125).

Schouten records that magnificent processions take place annually when the king goes forth, preceded by hundreds of elephants and their retainers, attended by musicians, sumptuously clothed nobles on horses and elephants 'richly adorned with precious Stones and Gold furniture'. These are followed by members of the king's guard, armed and richly dressed. Then seated in state on his mighty elephant comes the king himself, clad in his royal regalia, attended by lords and courtiers (Fig. 65). The heir to the kingdom follows, and, in ranking order, the king's wives and concubines, 'seated on elephants, in little enclosed Cabinets' (Figs. 66 and 67). Schouten estimates the procession at about 15,000–16,000 persons. Among these, surprisingly numerous were Japanese members of the king's élite guard. In fact, Japanese had already

Fig. 63
Informally attired European riding an elephant, with attendants, two of whom are turbaned Europeans in Persian dress. Detail from a modern facsimile replica of a pleated *khoi* paper manuscript copy of seventeenth-century murals at Wat Yom, Ayutthaya. (Photograph courtesy of the Fine Arts Department, Bangkok)

been present in Ayutthaya for some time as traders and had served as soldiers in the king's service. In 1632 they had been expelled for a short time for being troublesome, by King Prasat Thong, a stern legislator. During his reign, behaviour considered unseemly could lead to ingenious and unpleasant punishments: being sentenced to cut grass for the king's numerous elephants, surely a never-ending task, but better than the death sentence, that of being trampled to death by elephants.

Schouten also comments that the king's forces consist largely of unpaid conscripts, as well as 'Grandees' who in turn provide their own hundreds of attendants and elephants, so that the king's armies, at any one time, could range from 50 to 100,000 men (Caron and Schouten, 1671: 134). However, the army's strength largely derived from the hundreds of elephants, trained for war or to carry ordnances and provisions. Schouten's descriptions of the

Fig. 64
Demons and foreigners. By tradition
seventeenth–eighteenth-century foreign
residents of Ayutthaya are included in
depictions of the demon army. Detail of
mural at Wat Suwannaram, Thonburi,
mid-nineteenth century. (Photograph
Kim Retka)

hunting, taming, and training of elephants for this purpose reveals
that little has changed over the last few hundred years, as the
same methods continued into the early twentieth century (see
Chapter 7).

The European traders in Ayutthaya were a mixed lot. Bawdily
roistering, intriguing against each other, or diligently entering
their impressions in diaries or profits in account books, the traders
all vied for royal favour, for all trading monopolies and concessions were at the royal pleasure of the king and his ministers.
However, for some time, the growing trade strength and aggressive demands of the Dutch had caused the Siamese monarch
concern. By the mid- to late-seventeenth century, King Narai had
hoped to turn the European merchants' rivalry to the benefit of
Siam by playing the English against the Dutch. In this hope, he

Fig. 65
Engraving of courtiers and king on elephant-back. Clearly evident is the European engraver's obvious unfamiliarity with elephants and Ayutthaya architectural traditions. From Père Guy Tachard, *Voyage de Siam . . .*, 1686.

Fig. 66
Vessantara Jataka: Detail of royal procession, with ladies-in-waiting inside howdahs on elephants. Mural at Wat Suwannaram, Thonburi, mid-nineteenth century. (Photograph John Ringis)

Fig. 67
Re-enactment of a royal procession with warriors at the Surin Elephant Round-up. (Photograph Rita Ringis)

was encouraged by perhaps the most extraordinary European character in Siamese history, a Greek adventurer-turned-trader, Constans Phaulkon. From his runaway cabin-boy beginnings to humble employee of the English East India Company at Ayutthaya, Phaulkon had risen meteorically through opportunity, intelligence, and considerable linguistic abilities to the post of interpreter in the service of King Narai, and before long, to counsellor of the king in his dealings with foreigners, a position at the very pinnacle of the Court, the equivalent in power of a prime minister, a position also of great danger, enviously observed by jealous and increasingly xenophobic courtiers (Hutchinson, 1940).

However, as dealings with the English East India Company in this concern continued to be unsatisfactory to the Siamese, Phaulkon persuaded the king to turn to the French who were involved in extensive evangelical work for the Roman Catholic faith in the tolerant climate of Ayutthaya, seen then as a stepping-stone for further missions to China and Indo-China (see Plate 28).

At this time, Franco-Siamese relations were flourishing: the missions had expanded their good work, building churches and colleges, and French traders under the auspices of the French East India Company had arrived in Ayutthaya in 1680.

King Narai's admiration for the evident skills of the French as architects and engineers led him to arrange for them to design and build forts at Bangkok and Nonthaburi to protect Ayutthaya from possible attack from the sea. Although Ayutthaya is inland, some 60 kilometres by river from Bangkok, this fear of attack was a real concern, in a period of deteriorating relations between the Siamese and the Dutch (who had not been averse in the recent past to making their points understood by gunboat diplomacy). Thus, in turn, when the king decided to withdraw to Lopburi, deeming it safer than Ayutthaya, the French were also involved in the construction of the palace, fort, and other buildings (including extensive elephant stables), the magnificent but melancholy ruins of which may be visited today (Fine Arts Department, 1988).

Even now in their ruined (but recently partially restored) state, the fort and palace buildings of King Narai at Lopburi evoke a different world from that of Ayutthaya. The massive protective fort walls, with their numerous gate entrances, high enough for caparisoned elephants and their riders to pass through, are evocative reminders of the processions and cavalcades setting out for royal hunts. For there, freed from the formality of the capital, the king was able to indulge in his great love of elephant hunting in the forests that in those days surrounded Lopburi. It was here that a manual on elephant craft, thought to be the basis for subsequent treatises on elephant lore, may have been first compiled by King Narai, with the probable assistance of his then highly esteemed Commander of the Elephants, and hunting companion, Phra Phetracha.

The Siamese policy of tolerance and co-operation, as well as evident admiration of things French, may have misled the French into thinking that the king was also attracted to their religion. This line of thinking was probably fostered by the exchanges of embassies between Siam and the France of Louis XIV. In all, between the years 1681 and 1687, four embassies from Siam (the first was lost at sea) travelled to France, and two embassies came from France. Much pomp and gorgeous display blinded both sides to each other's aims: the Siamese wanted military expertise and assurances to offset the Dutch while the French took this as an agreement for French troops to be permanently garrisoned in Siam, and wanted in return nothing less than the conversion to Christianity of the Siamese king. It did not matter that this Buddhist king allowed all religions to coexist, and that he had received an embassy from Persia in 1685 without any indications of converting to Islam (Rong Syamananda, 1977: 75–80).

Inextricably caught between these conflicting aims by his own intrigues, vanity, and perhaps genuine desire to help Siam was Phaulkon as interpreter and go-between, but also, interestingly a very recent convert to Roman Catholicism and recipient of a timely knighthood from Louis XIV. In the event, things ended extremely badly for all. In 1688, as King Narai lay mortally ill, his protégé Phaulkon was accused of treason and beheaded on the orders of Phra Phetracha. As Commander of the Royal Regiment of Elephants, which were said to number 10,000, Phetracha was clearly the most powerful figure in the military hierarchy of Siam.

Prior to this event, contemporary descriptions of Phra Phetracha had noted his undoubted courage in battle, his lofty lineage, and the favour in which he was held, both by the people and the king, and that 'it is publickly reported that he or his Son ... may pretend to it [the throne] if either of them survive the King that now Reigns' (La Loubère, 1693: 89). As predicted, on the death of the king, Phra Phetracha swiftly disposed of King Narai's legitimate heirs and seized the throne, ridding the country of the threat of foreign domination and ushering in a long period of xenophobia.

Perhaps the most worthwhile and lasting result of the whole diplomatic débâcle was found in the writings by the French emissaries, invaluable descriptions of life in Siam under King Narai in the late seventeenth century. Though the most authoritative, exhaustive, and at times unintentionally amusingly ethnocentric account was produced by Simon de La Loubère of the Second Embassy during 1687–8, the Abbé de Choisy, accompanying Chevalier de Chaumont of the First Embassy (1685–6), gives a more personal account, in diary form. At times almost girlish in its sensibility, it focuses on the daily activities of the embassy and the grand receptions by the Siamese, recording the urbanity and courtesy of the king and his entourage towards the French emissaries in Ayutthaya, as well as in the less formal surrounds of the king's 'hunting lodge' at Lopburi.

The Abbé records with pleasure a journey of the entire French Embassy, all including himself on elephant-back, each elephant being directed by a man at the neck and one at the back, armed with goads. One cannot help but sympathize with his little aside that the ride was not comfortable but at least it was safe, the elephants being content to walk sedately. However, he notes that Phaulkon directs his own elephant sometimes at rather high speed. Being the eternal and perhaps prototype of the first tourist in Siam, the Abbé cannot resist mentioning that their promenade passes near a garden of the king, whose gardener happens to be French (Choisy, 1687: 57). Other promenades on elephant follow, grand processions with the king, excursions to view elephant duels, as well as visits to sites where wild elephants are being tamed, with Phaulkon bragging that in the kingdom there are 20,000 tame elephants, each with three men in attendance.

The Abbé was not unaware of feminine charms, and he wistfully describes the return of *la princesse-reine* from the hunt. He mourns that, alas, he will never see her, for before her elephant march overbearing guards obliging everybody to flee, and she, as her women also, rides in *une petite loge* or a covered howdah, for not even the courtiers may see her face.

Simon de La Loubère's work, on the other hand, is more object-ive and astonishingly observant and detailed, given that he spent only four months in the country. While recounting numerous details about the Siamese, their country, their customs, and their religion, La Loubère has much to say about the role of elephants in all aspects of Siamese life, whether that of commoner or king. He observes that in the palace there is always an elephant on guard, harnessed and ready to mount, tethered close to 'a little Scaffold, to which the King walks from his Apartment, and from this Scaffold he easily gets upon his Elephant' (La Loubère, 1693: 40). This type of structure can still be seen today in the palaces and temples of Bangkok where kings would have used them in much the same way as described by La Loubère (Fig. 68). These structures are in fact platforms, not free-standing but as part of a terrace or wall, set at the height of the shoulder of an elephant and to which the elephant would be tethered. Such arrangements enable a rider to step directly from the platform on to the ele-phant's back or into the howdah with some degree of grace and dignity. Of course, this method of alighting on an elephant or into a palanquin is used only by the king, who, as La Loubère points out, is thus never seen by the people actually walking on foot (except in his private apartments), thus enhancing his almost godlike status.

On rare ceremonial occasions, according to La Loubère, the king shows himself to the people and goes abroad in the city, either on elephant-back or by boat, and performs auspicious cere-monies in the main temples of Ayutthaya. La Loubère mentions specifically that the king never mounts the white elephant 'and the reason which they give is, that the white Elephant is as great a Lord as himself, because he has a King's soul like him' (La Loubère, 1693: 43).

La Loubère concludes that the elephant is also the principal domestic animal of the country, being the universal form of trans-port (apart from boats) and is thus hunted for that purpose by all, in the forests around Lopburi where it is plentiful. Describing in detail the capture and taming of wild elephants there, a method clearly used until very recently (see Chapter 7), he emphasizes that the animal is hunted for use, never for blood sports. Even court diversions and entertainments, such as a staged elephant combat witnessed by La Loubère, are carefully regulated as 'at Siam they neither expose the Life of Men nor Beasts, by way of Sport or Exercise' (La Loubère, 1693: 46).

While sceptically recording some of the beliefs that the Siamese held at that time about the elephant (for example, that it has a sense of justice, that it is as rational as a man, only lacking speech), La Loubère irascibly notes the difficulty of assessing accurately the numbers of elephants in the army 'by reason that

Fig. 68
Elephant-mount platform and
traditional tethering pillars at the front
of the Buddhaisawan Chapel in the
grounds of Wang Na Palace, National
Museum, Bangkok. (Photograph
Rita Ringis)

Vanity always inclines these People to Lying: and they are more
vain in the matter of Elephants, than in anything else' (La
Loubère, 1693: 89). However, he observes that the Siamese do feel
a deep affinity and affection for elephants, and that these animals
are more 'docible' than others. They surely must have been, to
allow three of themselves to be taken aboard the French ships
which would take them as gifts to the grandsons of Louis XIV. As
La Loubère notes,

The Siameses which brought them on Board our Ships to embark them,
took leave of them, as they would have done of three of their
Companions, and whisper'd them in their Ears, saying, Go, *depart cheer-
fully, you will be Slaves indeed, but you will be so to three* [of] *the greatest Princes of the World, whose Service is as moderate as it is glorious.*
They afterwards hoisted them into the Ships, and because they bow'd

83

down themselves to go under the Decks, they [the Siamese] cry'd out with admiration, as if all Animals did not as much to pass under low places. (La Loubère, 1693: 46.)

However, La Loubère had a very low opinion of the art of war in Siam, noting that the Siamese were 'little inclined to this Trade' (La Loubère, 1693: 90). This he attributed to a lack of courage, not specifically Siamese, but to all born in the 'excessive hot Countries', whether of European parentage or not. Moreover, he concluded, the Siamese belief in 'Metempsychosis' (transmigration of souls, or reincarnation) 'inspiring them with an horror of blood, deprives them likewise of the Spirit of War'. Thus, he notes with despair, when armies meet, they do not fire directly on one another but high above, in the hope that this will cause flight. This custom is also referred to by the Abbé de Choisy, but he had a different attitude to it: he felt that the Siamese fought 'comme les anges', and that to them, fear and retreat induced by the noise was in fact the main aim of a battle (Choisy, 1687: 92). Perhaps this accounts for the indecisiveness of traditional battles. Perhaps also, as La Loubère suggested, at the most basic levels of simple soldiers in battle, Buddhist teachings about the many paths of reincarnation thwarted the final blow to an enemy.

Apart from their comments on war, the observations of the French emissaries reflect a golden age, a time of openness and prosperity. During the thirty-two years of King Narai's rule, a united and strong Ayutthaya had prospered in commerce and flourished in peace. However, the dynasty founded by the usurper Phra Phetracha, while consolidating and building upon some of the fine traditions of the past, was none the less plagued by ineptitude and disunity, internecine strife, and generally speaking, poor leadership, contributing ultimately to the fall and destruction of Ayutthaya by the Burmese in 1767 (see also Fig. 8). After the torching of the city, and the looting of its treasures, tens of thousands of its citizens were force-marched to Burma, with many, including the king, dying on the way. No doubt, to transport the looted treasures to Burma, elephants in their hundreds were employed. Included ironically in the booty were large bronze Khmer sculptures captured by the Siamese during their own earlier campaigns in Cambodia. By a strange twist of fate, many Ayutthaya traditions were to be introduced and live on in Burma through the craftsmen captured in the fall of the city.[2]

After the 1767 destruction of Ayutthaya, to revive the Siamese state, a new kingdom was proclaimed in Thonburi. This lasted for fifteen years until its king, Taksin, was deposed because of his insanity. The throne was offered to his most able colleague, General Chao Phraya Chakri, who swiftly returned on elephant-

[2]As recorded in *Journey through Burma in 1936*, vestiges of those traditions were still evident in the early twentieth century. The Burmese called these 'Yo-Dayan', a term derived from Ayodhaya, a variation of Ayutthaya (Damrong Rajanubhab, 1991: 45).

back from his current campaigns in Cambodia and accepted the offer (Plate 22). Thus, in 1782, he became the founder of the Chakri Dynasty, of which the present monarch is the ninth king. (The kings of this dynasty are generally referred to by their regnal period and the name Rama. Thus the first king is known by foreigners as Rama I.)

For strategic and religious purposes, Rama I moved the capital from Thonburi across the Chao Phraya River in 1782 to the site of present-day Bangkok, then a mere marshy village. There he set about building a new capital, in accordance with ancient traditions. First and foremost was the construction of the City Pillar, the Royal Palace, and the Temple of the Emerald Buddha as the new focal point for all rites of the state and the monarchy (Fig. 69).

The Emerald Buddha image, actually carved from green jadeite, and thought to be of ancient origin, is the most venerated image in the land, being regarded by the Thai people as the palladium of the kingdom and the nation. The origin of the Emerald Buddha image is unknown, although many beliefs as to its divine construction and longevity have accrued over the centuries since it first appeared in Chiang Rai in 1434 in dramatic and seemingly miraculous circumstances which led to the spread of its fame, and subsequent veneration. In 1436, when the king of Chiang Mai sent for it, other miracles occurred. Mystifyingly, the elephant that was

Fig. 69
Exemplifying the best in Thai religious architecture, the Temple of the Emerald Buddha is the focal point of the nation's religious and state ceremonial. Enclosed by roofed galleries, on the far left is the ordination hall, and to the right are the Royal Pantheon, the Phra Mondop, the Phra Si Rattana Chedi, and the Ho Monthian Tham, the pediments of which feature Indra on the three-headed Erawan elephant. (Photograph Surachit Jamonman)

to transport the image in appropriately elaborate style from Chiang Rai to Chiang Mai thrice refused to take the road to its appointed destination, each time hastening instead into the town of Lampang, where the image was therefore allowed to be installed in a temple for the next thirty-two years (Fig. 70). In its subsequent travels during the next 300 years, the image was taken to Chiang Mai, to Luang Prabang, and to Vientiane in Laos, before being brought to Bangkok and then finally installed in the newly constructed temple in 1784 (Subhadradis Diskul, 1986: 18–19). It seems appropriate that today, over 500 years after the elephant-induced sojourn of the image in Lampang, an elephant obedience training school, famous throughout Thailand, is situated not far from the town.

Of necessity, the first three reigns of the Chakri Dynasty

Fig. 70
Commemorative statue of the elephant that carried the Emerald Buddha in the fifteenth century to Lampang at Wat Phra Keow Don Tao, Lampang. (Photograph John Ringis)

(1782–1851) were a time of rejuvenation and consolidation of Thai traditions in art and architecture, sacred literature, and law. Whatever could be salvaged from the destruction was put to use. Surviving craftsmen set to work, or taught others their traditional skills, to revive ancient traditions in a new world. The ruins of Ayutthaya were further 'ruined', as their ancient bricks and stones were brought down to Bangkok, to re-create the new civilization. For this, barges were used on the rivers and canals, as were elephants on land, to transport not only building materials, but also vast Buddha images from all over the country, great vestiges of the past glories of former Thai civilizations.

Following ancient traditions, to symbolically ensure peace and prosperity, Wat Suthat was constructed at the then centre of Bangkok, as an earthly model of the divine order of the Buddhist–Hindu universe (see Fig. 16). Protective sculptures of the god Indra as ruler, riding his three-headed celestial white elephant, Erawan, even today bestow their benign power from the pediments or gable-boards of Wat Suthat (see Plate 9).

Throughout the nineteenth century, within the rapidly expanding city, more mundane elephants, harnessed to gigantic loads, were hard at work, helping in the construction of major temples and palaces. In those former times, their keepers would daily take them down to the Chao Phraya River nearby, to bathe. One favourite bathing place, particularly for white elephants from the Grand Palace, is still known today as 'Tha Chang', or Elephant Jetty and is at present occupied by a modern busy wharf for river buses and commuter boats. It is also the place where tourists board their long-tail boats for the mandatory '*klong* tour', to view the vibrant and crowded life along the banks of the Chao Phraya River and its numerous *klong* or canals that in the past formed the criss-crossing and only 'roads' of the 'floating city' of Bangkok, any other roads then being mere muddy elephant tracks leading down to the river and *klong*. Across the latter, sturdy bridges, some of which still exist today, were built to accommodate the weight of the working elephant (Sirichai Narumitr, 1975).

Up-country, elephants were also deployed in the continuing protection of the realm, particularly from sporadic raids by the Burmese whose troops and *matériel* vastly outnumbered those of the Thai. Ever resourceful, the Thai resorted to many clever tricks to confuse the enemy. At one such engagement in 1785 in the Kanchanaburi area, near the Three Pagodas Pass, the Thai deceived and demoralized their enemy by simulating vast numbers of reinforcements to outnumber them. Day after day, scores of their elephants and men were seen to be passing into the Thai camp. At night, the same elephants and men would, under the cover of darkness, return to the base camp and at dawn, begin those journeys of 'reinforcements' again (Plu Luang, 1988: 36). A most notable ruse leading to victory is said to have occurred against enemy forces in the east during the mid-nineteenth century. This highly dramatic but possibly apocryphal tactic involved a surprise attack and understandable rout of the enemy in the darkness of night by a vast contingent of galloping elephants with flaming torches attached to their tails (Pallegoix, 1854).

By the mid-nineteenth century, with the British in control in Burma, the Thai were finally to be free of Burmese expansionist tendencies. Paradoxically, however, with this freedom from the Burmese threat came a new threat: the heightening of European colonial interests and imperial ambitions in the South-East Asian region. For Siam, this at first meant a concerted renewal or, in many cases, initiation of ties of friendship and informal commerce on a small scale with countries of the West. Portugal, Britain, and the United States were the earliest to tentatively seek diplomatic relations. Though formal treaties with Western countries were not signed until the reign of Rama IV in the mid-1850s, foreigners began to arrive in increasing numbers in a variety of capacities. Numbers of American merchants as well as Protestant medical missionaries had been arriving from 1828 onwards. The Roman Catholic French were not far behind. While the missionaries of either stamp were to have little success in converting the Siamese to Christianity, they were undeniably pioneers in the introduction of Western ideas. In this, they were to have, indirectly, an enormous influence on the country in the long term, through their cordial and long-standing scholarly relations with the scholar-monk who became in due course the progressive King Mongkut, or Rama IV (Fig. 71).

In the West, this king is known as the central character in the musical (and the movie), 'The King and I', based on the rather self-centred, occasionally informative, but frequently highly imaginative memoirs of Anna Leonowens, who was hired by the king as Governess to teach English to his numerous children in 1862. The king presented in Mrs Anna's book as something of a tyrant differs greatly from the king as known by his contemporaries and his legacy. King Mongkut's scholarly interests had led him beyond tradition-bound learning to wider pastures. Apart from his mother tongue, Thai, he was well versed in Pali and Sanskrit (the languages of Buddhist learning) and had some knowledge of Latin and Greek. But it was his knowledge of English (learnt from his friendly contacts with the missionaries) that provided the key to Western thought and scientific knowledge, allowing him a wide-ranging and realistic understanding of the outside world.

It could be said that this understanding was largely responsible for the fact that, uniquely of all South-East Asian countries, Siam was never colonized. King Mongkut realized the dangers of chauvinistic isolationism in the face of European colonial tendencies backed by superior armaments. In 1855, he welcomed the British envoy, Sir John Bowring, and in due course a Treaty of Friendship and Commerce was concluded between the two countries. Other treaties rapidly followed with the United States, France, Denmark, the Netherlands, Portugal, Prussia, Sweden, Norway, Belgium,

Fig. 71
His Majesty King Mongkut (Rama IV),
who reigned from 1851 to 1868.
Photograph from J. Thomson, *The Straits
of Malacca, Siam and Indo-China. . .,*
1875.

and Italy. Embassies were exchanged with Great Britain and France, these being the colonial powers between which Siam provided a wary buffer *vis-à-vis* their respective consuming interests in Burma and Indo-China.

In addition to his kingly and diplomatic duties, ecclesiastical reforms, scholarly pursuits, and studies in astronomy, King Mongkut found time to compile many interesting writings (Seni Pramoj and Kukrit Pramoj, 1987). These included his reflections on a subject dear to the hearts of the Siamese people: elephants and traditional beliefs about them, their classification into 'families', and their identifying characteristics for the good and ill fortune of their owners (Fine Arts Department, 1990). Many of these observations have been incorporated throughout this publication.

King Mongkut also conducted a wide correspondence in his distinctive and bookish English with many heads of state, thus establishing and maintaining cordial relations with their countries. Indeed, the king's correspondence with the British Consul in 1862 reveals his accurate and acute diplomatic concern about the possibility of misunderstandings between Siam, and France and England. As he had sent elephants to the Paris Zoo, 'it might appear that I am delighted to have much more respect and favourable endeavour for service to the Emperor of France than to Her Britannic Majesty who has entered the true friendship with me before the Emperor of France ...'. Arrangements were hastily made with the British Consul in 1862 to rectify this situation, and elephants were also offered to Queen Victoria (Seni Pramoj and Kukrit Pramoj, 1987: 158).

Possibly one of the king's most famous letters, originally directed to President James Buchanan of the United States (but answered in due course by the later elected President Abraham Lincoln), proposed sending pairs of male and female elephants which could be

turned loose in the forests where there was an abundance of water and grass in any region under the Sun's declination both North and South, called by the English the Torrid Zone—and all were forbidden to molest them; to attempt to raise them would be well, and if the climate there should prove favourable to elephants, we are of opinion that after a while they will increase till there be large herds as there are here on the Continent of Asia until the inhabitants of America will be able to catch them and tame them and use them as beasts of burden, making them of benefit to the country (Vimol Bhongbhibhat et al., 1987: 50–3).

With the benefit of present hindsight, it is clear that King Mongkut's eloquent and charming letter describes a state of affairs only possible in an ideal (past) world and his suggestion is a present-day elephant conservationist's dream. On the other hand, President Lincoln's cordial reply foreshadowed a future into which Siam was rapidly to move: modernization and thus Westernization on a grand scale in the coming century. Addressing the king as 'Great and Good Friend', Lincoln regretted that the United States did not have a climate 'as to favor the multiplication of the elephant, and steam on land, as well as on water, has been our best and most efficient agent of transportation in internal commerce' (Vimol Bhongbhibhat et al., 1987: 53).

It was during the reign of King Mongkut's son, Chulalongkorn or Rama V, who ruled from 1868 until 1910, that Siam progressed further along the path of modernization. The administration of the country was largely reorganized after Western models and in this the young king was assisted by some of his very able brothers, as well as resident Western advisers (Plate 23). As railways were constructed linking Bangkok with provincial outposts, 'steam on land' did indeed begin to replace the traditional mode of transport, the elephant.

King Chulalongkorn was the first Siamese king to travel abroad

Fig. 72
Bronze elephant memorial commemorating King Chulalongkorn's visit to Singapore in 1871 in the grounds of Singapore's Parliament House. (Photograph Rita Ringis)

since the sixteenth century (when Naresuan had gone to Burma as a child hostage). In 1872, he journeyed to Singapore, commemorating his visit with the presentation of a bronze statue of a Siamese elephant on a plinth pedestal which today stands in the grounds of Singapore's Parliament House (Fig. 72).[3] On that first voyage,

[3]Set in a leafy ground cover, the plinth base of this statue features inscriptions on its four faces in four languages, including Thai. The English inscription is as follows: 'His Majesty Somdech Phra Paramindr Maha Chulalongkorn, the Supreme King of Siam, Landed at Singapore. The First Foreign Land Visited by a Siamese Monarch on the 16th March 1871.' Most interestingly, there is a small well-trodden track leading through the ground cover to the face with the inscription in Thai, suggesting informal pilgrimages there, either by numerous Thai tourists or the Thai construction workers resident in Singapore.

91

King Chulalongkorn also visited Java, being most impressed by both places and the progress that had been achieved there under colonial rule. However, his own wide-ranging reforms were aimed at achieving that progress within a purely Thai framework. His subsequent two visits to Europe established very cordial relationships with European heads of state, and broadened his understanding of the West, most necessary and useful, given the still simmering colonial aspirations of England and France in the Indo-China region.

Whenever King Chulalongkorn travelled by his steam yacht, the *Maha Chakri*, unfurled from the mast was the flag of Siam, a white elephant on a red background (Plate 24). This design had been in use since 1817, when Rama II commemorated his auspicious white elephants in this way. It was this flag, seen on state occasions in Siam and abroad, that no doubt contributed to the country being known as 'The Land of the White Elephant'.

During the subsequent reigns of Rama VI and Rama VII, increasingly more Siamese travelled and studied abroad, a tradition begun in the previous reign when King Chulalongkorn had sent many of his sons (including the two who became the above-mentioned kings) to be educated abroad. The European-educated Thai intellectuals contributed to a new mood in the land, the desire for a more representative form of government. In 1932, a bloodless *coup d'état* ended the absolute monarchy, and a Constitution (the first of many) was promulgated. As the kingdom progresses on its sometimes tumultuous journey to a fully-fledged democracy, the constitutional monarchy continues to be held in the highest esteem by the Thai people.

5　The White Elephant

MYTHICAL creatures and exotic improbabilities of nature have always captivated and exercised man's fertile imagination: springing to mind are the dragon and the phoenix, both variously realized from ancient times in the mythologies and arts of the East and West. Other fantasies embrace the seductive mermaid, the mysterious yeti, the pious unicorn. While the mermaid has essentially lost her reputation and the yeti is, as always, almost on the verge of discovery, the British coat of arms still flaunts the literally fabulous unicorn, no less than an 'exquisite white pony, with a goat's beard, a flowing tail, and a long horn growing straight out of the middle of its forehead' (Clark, 1977: 37).

To that list of chimerical mysteries should be added the white elephant of Siam which embellished the country's flag until early this century (Fig. 73). However, unique among all the above dearly loved enigmas, the 'white' elephant was and is real, and respect for it is of long standing, going back well beyond the time of even the earliest European descriptions of elephants, which were by their definition, grey. Elephants of the everyday variety had become increasingly familiar to ordinary Europeans by the mid-nineteenth century through a then relatively recently introduced popular form of entertainment, the travelling circus. Then, as now, people were no doubt charmed and highly amused by the

Fig. 73
The White Elephant Flag. (Photograph courtesy of the National Museum Volunteers, Bangkok)

93

paradoxical amenability and docility of these huge and powerful beasts, performing unexpectedly nimble tricks. They could hardly have appreciated the fact that these endearing animals were fundamental to more serious pursuits: the overland transport of peoples and goods, as well as military security in some countries of the South-East Asian region. That the pale variations of these sturdy beasts were regarded as sacred would have been beyond comprehension to Europeans.

In the Thai language, the term 'white elephant' is *chang pheuak*, which literally means 'albino (or strange-coloured) elephant', the usual word for the colour 'white' being different entirely. To the Thai, encompassed in the term is an immediate recognition of the various subtle characterisitics that differentiate the white elephant from its more common grey cousins, ranging from the unusual number, colour, and shape of the elephant's toe-nails to the varying degrees of paleness of its eyes and skin. The perceived importance of these characteristics, traditionally deemed as auspicious, was actually formally acknowledged relatively recently when they were incorporated into Thai Law in the Elephant Conservation Act of 1921. In fact, the term immediately calls to mind the long association of the concept of the white or auspicious elephant with ancient traditions of religion and the monarchy, integral components in the fabric of Thai history and culture.

Though documented centuries before in ancient India, the first record in Thai of an actual white elephant appears on the thirteenth-century stone Inscription No 1. While elephants in general, whether as objects of commerce or aids in battle are frequently mentioned in this inscription, the singling out of a particular white elephant for special attention as a ceremonial vehicle implies even then its long-accepted status as sacred. The inscription records that customarily 'on the day of the new moon and the day of the full moon, when the white elephant named Rucasri has been decked out with howdah and tasselled cloth, and always with gold on both tusks, King Rama Gamhen [Ramkhamhaeng] mounts him' and journeys in stately procession to the forest monasteries (on the outskirts of the city of Sukhothai) to pay respects to the teachings of the Buddha (Griswold and Prasert na Nagara, 1971: 214–15).

History and tradition aside for the moment, in scientific terms, 'whiteness' or albinism in elephants is not hereditary. A 'white' or otherwise distinctively pale-hued elephant, almost always born in the wild to an 'ordinary' elephant, is merely a fortuitous conjunction of events and of genes, a 'slip' of nature, as it were, but a slip that became the symbol of a people. Embodied within the symbol was a view of the world at once ancient, yet entirely new to the European imagination.

In the nineteenth century, the horizons of the European general public were being considerably broadened and enriched by the continuing colonial thrust of Western nations into the Far East,

Thai elephants in the forest. (Photograph courtesy of the Tourism Authority of Thailand)

Elephants training to work as a team. (Photograph courtesy of the Forest Industry Organization)

4. Procession of royal ladies and child in a howdah on elephant-back. Detail of a mural at Wat Ko Keo Suttharam, Petchaburi, 1734. (Photograph John Piercis)

3. False door surmounted by a lintel featuring the god Indra riding the three-headed Eravana elephant at Prasat Hin Ban Phluang, Surin Province, eleventh century.

6. Lord Ganesha (Vighanessuan). Nineteenth-century manuscript. (Photograph courtesy of the National Library, Bangkok)

5. Statue of Ganesha in the grounds of the National School of Dance and Dramatic Arts, Bangkok, twentieth century. Visible on the building behind is Ganesha as emblem of the Fine Arts Department. (Photograph Rita Ringis)

8. The three trunks and tusks of the Erawan elephant, mounted by Indra, are visible in the axial niches at the upper body of the *prang* at Wat Arun, Bangkok. (Photograph

7. Pediment enclosing Dancing Siva and Ganesha (fragmented) at lower right. The lintel below depicts Reclining Vishnu. Eastern porch entrance at Prasat Khao Phnom Rung,

10. Pediment depicting Indra riding the three-headed Erawan elephant at Wat Rajburana, Bangkok, twentieth century. (Photograph Sarah McLean)

9. Detail of a pediment depicting Indra on the three-headed Erawan elephant at Wat Suthat, Bangkok. (Photograph Isabel Ringis)

11. Life of the Buddha: Detail of the drowning demon army in the Victory over Mara. Mural at Wat Bang Khanun, Thonburi. (Photograph John Ringis)

War elephant enactment at the Surin Elephant Round-up. (Photograph Rita Ringis)

. Re-enactment of a duel on elephant-back at the Surin Elephant Round-up. (Photograph courtesy of the Tourism Authority of Thailand)

14. Vessantara Jataka: Prince Vessantara meets four Brahmins who plead for his white elephant. Nineteenth-century manuscript. (Photograph courtesy of the National Museum Volunteers, Bangkok)

15. The white elephant as manifested by the Thirty-three Gods in Tavatimsa Heaven. Nineteenth-century manuscript. (Photograph courtesy of the National Library of Thailand)

5. Mahosadha Jataka: Princes on war elephants besieging a city. Detail of a mural at Wat Nayrong, Nonthaburi, nineteenth century. (Photograph John Ringis)

17. Elephants and mythical creatures in the Himaphan Forest. Mural in the ordination hall at Wat Suthat, Bangkok, early to mid-nineteenth century. (Photograph Kim Retka)

18. Detail of celestial multicoloured elephants.
    Mural at Wat Suthat, Bangkok.
    (Photograph Isabel Ringis)

19. Golden elephant with gem-studded filigree
    trappings, Chao Sam Phraya National
    Museum, Ayutthaya. (Photograph courtesy
    of the Fine Arts Department, Thailand)

20. Procession of elephants bearing 'castles' (howdahs) and riders. Mudmee silk textile, twentieth century. (Photograph Rita Ringis)

21. War elephant with two demon riders and attendants. Detail from the *Ramakian* murals in the galleries of the Temple of the Emerald Buddha. (Photograph Isabel Ringis)

. General Chao Phraya Chakri, founder of the Chakri Dynasty, accepting the throne on his arrival at the fort of Bangkok. (Photograph courtesy of the National Museum Volunteers, Bangkok)

. His Majestry King Chulalongkorn (Rama V) receiving a European delegation in a pavilion, while two royal tuskers and soldiers wait outside. Detail of a mural in the Royal Ordination Residence at Wat Benchamabophit (Marble Temple), Bangkok, late nineteenth century. (Photograph Elizabeth Dhe)

24. His Majesty King Chulalongkorn (Rama V) in European military uniform reviewing troops. In the foreground is the Royal Steam Yacht *Maha Chakri* and White Elephant flags.

Mural at Wat Chanasongkram, Bangkok, twentieth century. (Photograph John Ringis)

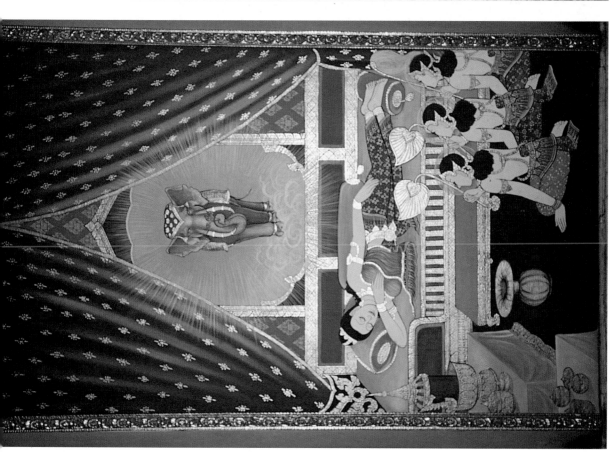

26. Eighteenth-century scripture cabinet door panels in black and gold lacquerware depicting the *Traibhum* with Mount Meru at the centre and Anotatta Lake and elephants at the lower left, Chao Sam Phraya National Museum, Ayutthaya. (Photograph Eileen Deeley)

28. The Buddha taming the Nalagiri elephant. Below is a vignette of Europeans, one of them in a Thai Buddhist monk's robe. Detail of a mural at Wat Ko Keo Suttharam,

27. Gilded wood carving on the door panel at a rural temple depicting the Buddha's years in the Palileyyaka Forest with animals of the forest offering him sustenance.

and the resultant growing international trade and travel in that region. While traders and merchants concentrated on economic gain, gentlemen travellers of the time penned their memoirs, which received a wide readership among the armchair travellers of the growing middle classes in the West. A notable object of the travellers' curiosity was the emblem chosen by Siam for its flag, the white elephant on a scarlet or blue ground, familiar to the Europeans from the early nineteenth century onwards on ships plying the oceans, and popularizing Siam as 'The Land of the White Elephant'.

Though references to the existence of white elephants in the region had been made by European observers as early as the sixteenth century, it is in general these nineteenth-century accounts that not infrequently helped to perpetuate the European sense of superiority, widening the vast gap that seemed to separate the apparently enlightened civilizations of the West and the so-called benighted realms of the East. Sensationalism in reporting is not confined to the twentieth century.

One such sensationally inclined nineteenth-century traveller was F. A. Neale (prolific writer on the exotic, listed on his title pages as 'Author of Eight Years in Syria, &c'). In his *Narrative of a Residence in Siam*, originally published in 1852, Neale expressed ill-informed though no doubt 'democratically' motivated outrage on a purported visit to a 'Temple of a White Elephant' (no such place ever existed). There, Neale (1852: 98) apparently observed these 'most revered of all the Siamese deities ... [the image of] which ... floats proudly for the Siamese in the banner of their nation. An elephant is certainly more terribly emblematic of tyranny than anything I know of; at least in my humble opinion.'

Admitted into the presence of 'the brute', Neale records, *unlike* any observer before or after, that 'his skin was as smooth and spotless and white as the driven snow'. According to Neale (1852: 99–100), though the actual room housing the elephant was 'unpresuming', the flooring was covered with a mat-work, wrought 'of pure chased gold' prompting him to exclaim 'if this was not *sin to snakes*, as the Yankees say, I don't know what was. The idea of a great unwieldy brute like the elephant, trampling underfoot and wearing out more gold in one year than many hardworking people gain in ten!'

Such observations and out of context comments of the extraordinary treatment of the white elephant, a mere animal, and the luxury in which it was maintained, were bound to astonish Europeans. No doubt, outraged comments such as these ultimately contributed to Western misunderstandings and misconceptions about the white elephant being enshrined as a derogatory term in English. Every English language speaker, though he or she may not have the slightest notion of where 'Siam' is or was, will know and readily use the term 'white elephant'. Sacred in Siam, in English it has become a term of abuse, varying only in the degree of bemused or affectionate contempt expressed for an object thus described. For example, on a small scale, a white elephant stall in a village or community fête aims to sell odd and useless objects, donated by their hopeful owners for charity; on a larger scale, an unpopular and expensive project by a government is a 'white elephant' and is sure to be castigated as such.

The *Shorter Oxford English Dictionary* defines a 'white elephant' as 'a burdensome or costly possession (given by the king of Siam to obnoxious courtiers in order to ruin them)'. But no Siamese monarch ever considered white elephants 'burdensome' nor gave them away, for according to ancient tradition, possession of one or many of these symbolized a king's virtue or *barami*. Indeed, in the mid-sixteenth century, the refusal by the Siamese Crown to give away 'superfluous' white elephants resulted in a bitter war between the Siamese and Burmese.

Perhaps the derogatory English term arose from a misapplication to Siam of the comments made by various European observers, from the sixteenth century onwards, about traditions of maintenance of the white elephant in Burma. These comments were consolidated and possibly given currency by the 1858 report of the British mission to Burma led by Henry Yule in 1855, and later by the American traveller, Frank Vincent, in his engaging and then highly popular narrative of his journey through *The Land of the White Elephant: Sights and Scenes in South-East Asia, 1871–1872*, originally published in 1874, but running to several later editions, testifying to the expanding market for exotic tales for the armchair traveller of the time. These commentaries noted the heavy financial burden placed on the Burmese king's subjects for cost of feeding and maintenence of the white elephant, who would be assigned the income of provinces and cities to 'eat'. However, in European accounts of the white elephant in Siam, this practice is never alluded to.

Although the everyday elephant was indigenous in the South-East Asian region from prehistoric times, the concept of its sacred nature, in the form of a white elephant, is apparently more recent, dating back to Indian religious beliefs and traditions, both Hindu and Buddhist, which penetrated and influenced the indigenous cultures of parts of South-East Asia continuously from nearly 2,000 years ago. While Buddhism took root and flourished in the region today known as Thailand, the Hindu gods and their frequently rather lively companions were not rejected, but incorporated into a sort of pantheon of guardians. Stories of their exploits provided much of the basis of a sacred (as well as entertaining) literature. This was accessible not necessarily only through the written word (then only available to an élite) but to all who could see and judge: in the art and architecture inspired by religious beliefs throughout the region. Architecture, particularly in the form of extensive temple structures, embodied symbolically various ancient concepts about the order of the universe, and the place within it of man, gods, and

96

various celestial creatures, including not the least, the white elephant.

Today, in Thailand, perhaps the best known and unexpectedly familiar manifestation of these complex traditional concepts is to be seen at the Temple of the Dawn, Wat Arun, an important place of pilgrimage for the Buddhist devotee, and a mandatory tourist destination for every visitor to Bangkok. Though the Temple of the Dawn was completed as recently as the nineteenth century, its symbolism reveals the persistence of ancient traditions in the land. Adorned with myriad multicoloured ceramic shards, the lofty central *prang* or tower represents the Hindu and Buddhist axis of the universe, the magic Mount Meru, at the summit of which is Tavatimsa, the Heaven of the Thirty-three Gods, chief of whom is Indra, Lord of Storms and War (see Plate 8).

Here, at Wat Arun, as earth-bound devotees and tourists gaze upwards, clearly visible high above, emerging from niches at the four cardinal points of the soaring *prang* are sculptural representations of the regally clad god Indra, seated on his traditional mount or vehicle, the celestial white elephant, depicted with three heads and thus three massive sinuous trunks to emphasize its power. It is this celestial elephant, called Erawan in Thai (and sometimes depicted with thirty-three heads) that is partly considered to be the prototype of the white elephant on earth (Fig. 74). According

Fig. 74
The Erawan elephant depicted with thirty-three heads, and pavilions with Indra and celestial attendants. Nineteenth-century manuscript. (Photograph courtesy of the National Library of Thailand)

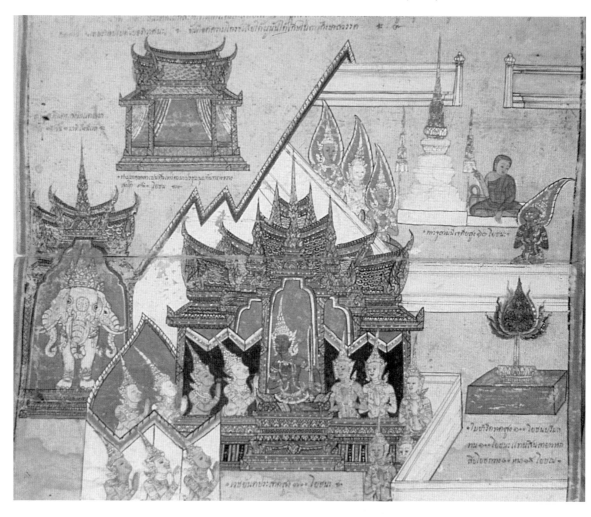

Fig. 75
Tavatimsa Heaven, with Indra in his palace, and Erawan in a pavilion. Detail from the *khoi* paper manuscript of the *Thonburi Traibhum*, late eighteenth century. (Photograph courtesy of the National Museum Volunteers, Bangkok)

to ancient beliefs and writings, just as Indra on his mighty Erawan maintained harmony in the celestial spheres, so on earth the possessor of the terrestrial manifestation of Erawan, in the form of a white elephant, would also be seen to have godlike powers within his own realm or universe (Fig. 75). This belief, though having its roots in ancient India, was certainly fostered in Siam by a seminal text considered today to be one of the earliest and finest works of Thai literature, commonly known as the *Traibhumikatha*.[1]

[1]Formally referred to as *Traiphum Phra Ruang*, the *Traibhumikatha* is written as a sermon, and draws heavily on concepts from the Pali Tripitaka Canon, the most authoritative text of Theravada Buddhism, compiled in Sri Lanka in the first century BC. Currently, there are two excellent translations of the Thai text into English: *Three Worlds According to King Ruang: A Thai Buddhist Cosmology* by Frank E. Reynolds and Mani B. Reynolds (1982) and *Traibhumikatha: The Story of the Three Planes of Existence*, by King Lithai, translated by the Thai National Team under the leadership of Khunying Kullasap Gesmankit and published in the Anthology of ASEAN Literature series (1987). While the translations do not differ largely in spirit or substance, the quotations that follow in this discussion are from

98

Believed to have been compiled in the mid-fourteenth century, though no original version of that time still exists, it appears to have been well known by the mid-seventeenth century, as the contents are clearly alluded to (though not by name, but subject) in the report of the French envoy to the kingdom of Siam at that time (La Loubère, 1693: 149). Extant eighteenth- and nineteenth-century versions of the *Traibhumikatha* text describe graphically and poetically the Three Worlds (comprising thirty-three planes or levels of existence) into which are bound to be reborn all manner of beings, including animals, men, and even the gods, according to their merits.

These planes encompass infernal, earthly, and heavenly regions. Within the human realm are born not only ordinary men, but from time to time, beings of the greatest accumulated merit. According to the text, if such an individual appears within any given era, he would be considered as Chakravartin, the Universal Emperor, or the righteous Buddhist king. His legitimacy is proclaimed by his possession of seven miraculous attributes which accrue to him 'spontaneously' if he zealously pursues the ten principles of righteous kingship. Thus any worldly ruler possessing these attributes, which included white elephants, would be recognized by his people as such a righteous monarch, whose celestial virtues are poetically enumerated in the text. Many a Siamese (as well as Burmese) king was to model himself on such an exalted ruler, distinguished by the title of 'Lord of the White Elephant'. Ironically, sporadic and bitter wars between the two regions for several centuries can be partially accounted for because of conflicting claims to this title, as only one Chakravartin can exist in any given era.

Although most effusively and colourfully described in the text, many of these apparent celestial attributes are logical and useful even to the most mundane ruler, for they imply political stability in any age: a Perfect Woman (the queen), the Precious Chamberlain (or adviser), and the Excellent General. Supernatural attributes include the Bejewelled Wheel and the Precious Gem, the constant presence of which enhance and brighten the ruler's journeys throughout his universe. Above all, to speed the Universal Emperor on his way is the Precious Horse, and most importantly (to this subject anyway), the Most Noble of Elephants.

According to the text, this Noble or Precious Elephant enables the righteous Chakravartin to ride throughout his universe 'like the Lord Indra mounted on Aiyarawana [Erawan] elephant'. The *Traibhumikatha* text enumerates the characteristics of such an elephant, which is a member of a family of celestial elephants

the more recent 1987 publication which, most usefully for the interested reader, presents the text in parallel form, both in Thai and English. However, given the wider availability of the Reynolds translation, page references for that publication are given as well for the interested reader's convenience.

99

called 'Chaddanta'. Elephants of that family are 'unsurpassed in their strength and beauty' (see Plates 17 and 18). They are 'white of colour, like the glow of the moon in its full glory. The soles of their feet are rosy like the sun at first dawning ... their trunks are as red as the lotus' bloom ... they are possessed of supernatural powers ... [and] they ... are as magnificent as silver mountains....' (*Traibhumikatha*, 1987: 201; Reynolds and Reynolds, 1982: 161.)

The text also cites in detail the honours accorded in the heavenly spheres to such incomparable elephants: for example, the preparation of a befitting residence, a pavilion supported by gold and silver pillars, 'decorated with gems of seven kinds', and as wonderful as 'the celestial abodes of the devyata [gods]' (*Traibhumikatha*, 1987: 199; Reynolds and Reynolds, 1982: 160). Aromatic pastes and fragrances perfume the inner area where, canopied by fine cloth embroidered with gems, a golden dais awaits the miraculous arrival of the Noble Elephant which renders homage to the Chakravartin by spontaneously bowing until its tusks touch the ground. Then, by royal command, this elephant is adorned with

fineries and trappings of great worth, ornaments of gold and silver, and gems and jewels and priceless fabrics.... Then the King mounted the Precious Elephant ... [who] bore him up swiftly into the sky.... How glorious the Chakravarti King seemed amidst his host of retinue; how like the Lord Indra mounted on Aiyarawana elephant surrounded by the devyata [celestials].

In his celestial journey, the elephant's speed is such that within the blinking of an eye 'the King and his retinue circled Mount (Su)meru and followed the wall of the universe around in a full circle' (*Traibhumikatha*, 1987: 203; Reynolds and Reynolds, 1982: 161).

Here in this Thai Buddhist text, traditionally accepted to have been originally compiled and written in the fourteenth century, is reiterated and reaffirmed the importance of much older and originally Hindu concepts of Mount Meru as the axis of the universe, and the ancient supremacy there of the god Indra and his white elephant. All three concepts are embodied in various ways in the decoration, design, and layout of Buddhist temples in Thailand, whether of the remote past or present day (see Chapter 6).

The poetry of the text had a practical purpose as well, for the lyrical descriptions of celestial practices provided a model, an etiquette as it were, for a righteous ruler's behaviour on earth, and not least of all, for the treatment of terrestrial white elephants as befitted their dignity. Seen in this context, the customs of building sumptuous pavilions and celebrating elaborate rites for the reception and induction of a white elephant on earth are not, as was certainly later perceived by European observers, the frivolous waste of resources on a mere animal. Indeed, such ceremonial represented a ritual re-creation and re-enactment in miniature of a vision of the divine, and thus was regarded as visible confirmation of an earthly ruler's righteousness to wield power for the benefit of his people.

The text of the *Traibhumikatha* continues, most sensibly, to note that 'even though the blessed Chakravarti King has subdued all four parts of the world, he is still a mortal and is impermanent for he still wanders in the cycle of birth and rebirth'. On his death, all the precious attributes leave him; even the Perfect Woman 'will lose her radiant glow and live as any ordinary woman'. Thus also will the Noblest Elephant return to his celestial regions and blessed family (*Traibhumikatha*, 1987: 221; Reynolds and Reynolds, 1982: 170–1).

This 'desertion' by the Noblest of Elephants, one of the seven attributes of the Chakravartin, helps explain the reasons for individual Siamese kings actively seeking, in the past, the capture of such white elephants. Symbols of legitimacy cannot be inherited, but must accrue 'spontaneously' in each reign. Indeed, not every Siamese ruler possessed white elephants. Thus, while in the West a white elephant was perceived as just a highly indulged mere elephant or pale pachyderm, in realms that differed from those of the West, specifically, in Siam, the possession of a white elephant was perceived as a sacred sign of celestial approval of the earthly state and its ruler. Here it may be pertinent to note that over the centuries, many of the European observers who turned up their collective noses at such curious symbols were themselves descendants of societies in which the Divine Right of Kings had been enshrined.

Despite common nineteenth-century misconceptions, Siamese beliefs about the sanctity of white elephants had already been noted in some detail from the late sixteenth and seventeenth centuries onwards by well-established European merchants in South-East Asia in general, as well as travellers and adventurers to the capital of Siam, Ayutthaya specifically. They recounted tales of the exotic customs of the people there and enumerated the equally exotic flora and fauna to be found in these regions. Invariably included among these were descriptions of the appearance of that mystery of nature, the white elephant, and the reverence and respect in which this creature was held by the people and their rulers.

In the seventeenth century, a Dutch trader, Van Vliet, recounted that the Siamese and neighbouring nations honoured the white elephant 'as a prince of the elephants', and that the king usually had at least one at court, 'which is well lodged, ornamented, well treated and provided with attendants and a suite; the king often pays it a visit, and it is fed from plates of pure gold'. Van Vliet further recorded that the white elephant was appreciated not only for its colour but 'also for its natural intelligence and for its sensitiveness for honour and treatment. It becomes sad and melancholy when not properly treated.... This seems fabulous but long experience has taught the Siamese to believe it.' (Van Vliet, 1692: 100.)

Other details of white elephants and their characteristics found their way to Europe in the comprehensive report on Siam in 1687–8 by the French emissary Simon de La Loubère, who pointed out that although

these Animals are rare ... they are not altogether White, but of a flesh colour.... The Siamese do call this colour *Peuak*, and I doubt not that it is this colour, inclining to White, and moreover so rare in this Animal, which has procured the Veneration of those People to such a degree as to persuade them ... that a Soul of some Prince is always lodged in the body of a White Elephant, whether Male or Female it matters not. (La Loubère, 1693: 98.)

La Loubère also noted that although the king (Narai) rode elephants, he *never* rode a White Elephant, as he deemed it to be possessed of 'a King's soul'.[2]

The acquisition of such an elephant, said La Loubère, was in itself held miraculous, as he had been told that they were not born in captivity, but were found in mysterious ways in the wild. Interestingly, this belief still holds strong in the twentieth century. Keepers at the present-day Royal Elephant Stables at Chitralada Palace noted that while generally elephants in the wild tended to stay close to their herd, a white elephant was always found on its own, having apparently wandered away from its herd towards the haunts of man, so as to fulfil its destiny, some would say. Others suggested that an elephant's albino and therefore out of the ordinary appearance might have something to do with its solitary wandering. Periodic discoveries of heavily mud-covered baby (white) elephants ('alone and palely loitering'?) implied that their mothers had tried to protect, in vain, their 'unfortunately' different offspring from being possibly ostracized by the rest of the herd. Whether such an elephant is driven away by the herd because of perceived physical differences or wanders away of its own accord because of its supramundane nature remains in the eyes of the beholder.

However, in the seventeenth century, Simon de La Loubère correctly observed that the Siamese were convinced of the white elephant's sanctity, their conviction being strengthened by sacred texts. Although at the time, La Loubère complained that he did not have access to the sacred texts themselves (La Loubère, 1693: 136) but had to make do with halting translations, through the dark glass of his at times somewhat unsympathetic descriptions, it is clearly evident from present hindsight that his Siamese companions were narrating 'Stories of the Former Lives of the Buddha' (the Jataka Tales), and the Life of the Buddha. According to Thai beliefs, in both of these traditions, white elephants feature in a variety of important ways. For example, in one of the Jataka Tales

[2]This seventeenth-century observation differs from that given on the thirteenth-century Ramkhamhaeng Stele which describes the king as customarily riding a white elephant. Scholars who doubt the authenticity of the stele view this contradiction by an objective foreign observer as another contributory fact to their argument (Piriya Krairiksh, 1991a: 104).

which illustrate virtues such as compassion and generosity in parable form, the Buddha-to-be, as the central character, takes the form of a white elephant in the Chaddanta Jataka. In another of the Tales, the giving away of a white elephant precipitates a series of events known as the Vessantara Jataka, for centuries a great favourite of the Thai people (see Plate 14).

Most importantly, the white elephant features in the story of the Life of the Buddha, and mural paintings in many Thai temples, both ancient and contemporary, illustrate that on the night of his conception, his mother-to-be, Queen Mahamaya, dreams that a white elephant approaches her and touches her right side with a lotus held aloft in its trunk, entering her womb (Plate 25). Naturally, over many centuries, such religious traditions contributed to the perception of the white elephant as sacred, and as a harbinger of good fortune. In fact, fused in the Siamese reverence for the white elephant were traditions fundamental to the culture: the practice of Buddhism, and as discussed earlier, the legitimacy of righteous kings.

Seen in this religious and political context, an extraordinary account by Van Vliet becomes somewhat more understandable. Van Vliet recorded that in 1633, because a young white elephant in the care of the court had suddenly died, the desolate king had ordered the execution of all the attendants of the late white elephant. In addition, a great funeral and cremation 'even greater than that which ever has been displayed for the most famous mandarins' was arranged for the dead elephant, this because 'the Siamese pretend that, besides royal dignity, there is something divine in these animals' (Van Vliet, 1692: 101).

From modern hindsight, one could surmise that both the execution of the attendants and the lavish funeral of the elephant were most certainly inspired by the fact that the king in question, Prasat Thong, had recently usurped the throne, and white elephants as symbols of legitimacy of a ruler, of the Chakravartin, would have been of utmost importance to the stability of his reign. Also clarified is the apparent 'foolishness' of the Siamese and Burmese in going to war during the mid-sixteenth century over the possession of white elephants (see Chapter 4). Every European account of the history of Siam wryly notes this, with heavy (and arch) implications as to the absurd frivolity of such a war. But seen in its proper historical and political context, it is no more absurd than the busy burnings at the stake, and varied religious persecutions current during the same time in Europe involving Lutherans, Protestants, Huguenots, Calvinists, Roman Catholics, and 'Papists' of various stamps.

In the early nineteenth century, Siamese traditional beliefs were further reaffirmed when Rama II, then in happy possession of several white elephants, and himself an authority on elephants in general, devised the state flag bearing the white elephant encircled by the *chakra* or wheel symbol of the Chakri Dynasty on a scarlet background as a symbol of Siam, newly risen from the ashes of Ayutthaya and restored in the foundation of Bangkok in 1782.[3]

By the mid-nineteenth century, many European observers—minor royalties, professional travellers, scallywags, and officials in various capacities—had taken the opportunity to visit the exotic East, and among other things, to observe and pass judgement on that object of persistent curiosity in the West, the white elephant in the new capital of Siam, Bangkok.

In contrast to some of the more sensational reports, one early nineteenth-century observer who has left us a reasonably object-ive scientific description was George Finlayson, a surgeon and nat-uralist accompanying the East India Company Mission to Siam and Hue, led by Dr John Crawfurd, in 1822. In Bangkok, Finlayson visited the Siamese king's white elephants and recorded in his report that these albino creatures differed only subtly from their more mundane counterparts, their skin being fairer, their hairs being yellowish, but most interestingly, the irises of their eyes being a pure white colour. As a scientist, he commented that these eyes, unlike those of a human albino, were not intolerant of light, but perfectly sound (as is the case apparently with other quadruped albinos). Free from the prissy distate that was to colour some traveller's tales, he noted that the animals were at that time in excellent condition, kept on a small platform for cleanliness, and fed abundantly with fresh grasses, sugar-canes, and plantain bunches (Finlayson, 1826: 152).

Another reliable late nineteenth-century witness was the indefati-gable and perceptive traveller, Carl Bock, author of *Temples and Elephants: Travels in Siam in 1881–1882*, originally published in 1884. Confirming the respect in which white elephants were still held in the kingdom, he also noted that in charge of all arrange-ments concerning the royal elephants was a noted authority on them, a most important and distinguished man, in fact the king's uncle, Prince Maha Mala. This hereditary position of honour, Lord of the Elephants, was official until recent times and is still a source of pride to twentieth-century descendants of the prince.

In his writings, Bock, as had many others before him, confessed to a slight disappointment at finding that the white elephant was not as he had expected, really white, but of

---

[3]In time, the wheel symbol was omitted, leaving the white elephant silhouetted on the scarlet. The White Elephant Flag was superseded only in 1917, when the Thai sent a contingent during the First World War to the Allies in France. Various reasons have been surmised for the change in the flag. While on the one hand there was reluctance to see the potential trampling of the white elephant symbol on a fallen flag in the field of battle, there was also the feeling that 'the red background of the flag boded ill for a monarchy in view of the fact that the Bolsheviks were at that time using a red flag' (Rong Syamananda, 1977: 2). Most probably, however, given the Thai genius for adaptation, the change resulted from a wish to present a modern face to the world. Since 1917, the flag has consisted of five horizontal bands of red, white, blue, white, and red.

a pale reddish brown colour, with a few real white hairs on the back. The iris of the eye, the colour of which is held to be a good test of an albino, was a pale Naples yellow. He looked peaceful enough ... and his quiet bearing was in contrast with the excitement all around, as if he felt the importance of his position. (Bock, 1884: 25.)

Bock also recounted an interesting incident that reflected the depth of European misunderstanding about the white elephants of Siam. An English travelling circus, Wilson's, had visited Bangkok in 1881 and widely advertised to the Siamese populace a performance by a 'white elephant'. During the show, to the jesting and prompting of clowns, an elephant of snow-white complexion performed humiliating and belittling tasks, and as he did, not unexpectedly, his 'whiteness' left a white mark on everything he touched. Bock noted that all this was viewed by the Siamese in utter silence and obvious distaste. Agreeing that the whole performance was in bad taste, he recorded that the Siamese audience's conviction that this mockery would be vindicated was to come true in a short time. On the further journey of the circus to perform in Singapore, the elephant in question died, as did Mr Wilson, the proprietor of the circus (Bock, 1884: 33).

However, there was no denying the pomp and ceremony that had surrounded the *genuine* white elephants of their Siamese Majesties for centuries. To the Siamese, the ceremonial was not a frivolous display but an act of homage to the earthly representative of the gods and guardians of the universe.

Even King Mongkut, or Rama IV, nowadays acknowledged as the astute and enlightened monarch that he was (as against the temperamental autocratic tyrant presented by Anna Leonowens, the English governess), was enchanted by the enigmatic beauty of the white elephant. In late 1854, in the king's correspondence with Sir John Bowring, Ambassador Plenipotentiary and leader of the British Government Mission to Siam, His Majesty could not resist mentioning his pleasure at the capture of 'a whitest she Elephant ... which we say to be admirable for adorn[ment] of our city. The greatest ceremony will be taken here on her arrival according to the former custom upon the reigns of my royal grandfather and late father, when their whitest Elephants were arrived.' (Seni Pramoj and Kukrit Pramoj, 1987: 104.)

Such ceremonies were times of rejoicing for all. Over the centuries, many observers have recorded generally similar impressions, attesting to the longevity and continuity of these Siamese traditions. If present-day ceremonial barge processions held on important state occasions are anything to go by, then these welcomes for the white elephant were certainly grand. The invariably rustic captors of the elephant were amply rewarded, even ennobled in some cases (thus no doubt contributing to the diligence with which these animals were sought). The general populace, dressed in their brightest colours, gathered from far and wide, filling the flag- and bunting-decorated city and lining the banks of the great Chao Phraya River to welcome the flotilla of glittering, fantastically carved barges whose hundreds of oarsmen plied their gilded oars with clockwork precision, kept in perfect time by traditional boating chants, strangely melancholy to the European ear (Fig. 76). At the centre of this gliding procession was the sumptuously canopied and flower-garlanded raft bearing the 'new' white elephant.

Prior to the reign of King Mongkut in the mid-nineteenth century, when Siamese kings travelled, the people had been forbidden to view their passing 'lest they should desecrate His Majesty by their unclean glances' (Manich Jumsai, 1991: 62). This custom had been introduced centuries before during the Ayutthaya period when the Siamese kings had adopted Khmer traditions of according the monarch god-king status. However, King Mongkut, who had travelled extensively among the people during his twenty-seven years as a scholar-monk prior to his accession, abolished this practice and appeared frequently in public. Thus, since the mid-nineteenth century, on occasion, the king had accompanied the river processions or welcomed the elephant on shore, travelling to the disembarkation point in regal procession, perhaps carried on a gilded palanquin shaded by a huge golden parasol. Also wending their way to the river were processions of richly clad nobles attended by their retainers bearing precious vessels and other insignia of rank. Following them were soldiers in striking uniforms, and a cavalcade of the state elephants plodding sedately, encumbered by gold trappings and intricately wrought and canopied howdahs, in which were seated princesses or younger

Fig. 76
A royal barge on the Chao Phraya River. From Anna Leonowens, *The English Governess at the Siamese Court*, 1870.

members of the royal family. Though the occasion was one of
rejoicing, cheering was not customary then (nor now) in Siam on
state occasions, and the people maintained a hushed silence.
However, scarlet-clad musicians filled the air with their drumming
and the eerie high-pitched wail of ceremonial conch shells, joyful
sounds to the Siamese, but 'discordant' to the European ear, as
Bock (1884: 23) observed in such a procession in 1882.

After disembarkation, the entire entourage would proceed to a
splendid pavilion, by tradition temporary (the word 'temporary'
should not be construed to imply 'make-shift', for the Thai people
excel in the construction of exquisite ceremonial structures the
beauty of which is enhanced by the notion of their ephemerality).
At the pavilion, traditional court astrologers and masters of
complex and esoteric ceremonies, the white-clad royal Brahmin
(Hindu) priests, would officiate at the welcome. (Although the
saffron-robed monks of Buddhism represent the major religion in
the land, the Brahmins even today reflect the ancient Hindu heritage of Thailand, and conduct ceremonial and rituals that have
no place in Buddhism, focused as it is on individual progress
towards enlightenment.)

At the elephant's welcoming ceremony, the Brahmins invoked
'the angels who preside over the destiny of all elephants ... to
assemble now, in order that you may prevent all evil to His
Majesty the King of Siam, and also to this magnificent elephant,
which has recently arrived' (Young, 1898: 392). Turning their
attention to the sometimes bewildered and reluctant elephant, the
Brahmins would chant instructions to assuage its fear and soothe
it, begging the elephant not to think too much of its mother and
father, relatives and friends left behind in the forest, where evil
spirits lurk and wild beasts howl, and where the dreaded (mythical) *hastilinga* bird preys on elephant flesh (see also Fig. 35).
Calling upon the elephant to forget the forest, and its flies and
mosquitoes, its dust and danger, the Brahmins would chant assurances to the elephant that its life in the capital would be all that it
could desire for 'it is of your own merit that you have come to
behold this beautiful city, to enjoy its wealth, and to be the
favourite guest of His Most Exalted Majesty the King' (Young,
1898: 393). Awaiting the elephant was a life of protected ease,
attended by assiduous retainers who would bathe, feed, and care
for its every comfort. Its duties in return were not unduly onerous:
to be present, gorgeously arrayed, at all state occasions.

Such invocations also took place at the blessing and naming ceremony, in the presence of the court and dignitaries when the elephant
was bathed and 'baptized' by the Brahmins with lustral water. At
that time the rank of prince was bestowed upon it, as well as an auspicious name eulogizing the elephant's numerous virtues. As the
name was chanted in its lengthy entirety, the elephant would be
offered a stick of sugar-cane on which those same virtues were
inscribed, and which it invariably consumed with some relish.

In 1927, during one of the last grand nation-wide ceremonial receptions of a white elephant, ancient tradition coalesced with the more utilitarian modern age. A white baby elephant, born (unusually) in captivity within a herd belonging to the English forestry concession of the Borneo Company in the north of Thailand, travelled to the capital with due pomp and ceremony. However, the journey was not by traditional river procession but by the relatively recently completed rail link between Chiang Mai and Bangkok, with stops at many stations to allow for regional welcoming ceremonies, lengthening the (then) usual 24-hour duration of the journey to seven days before arrival in Bangkok, and further ceremonial (Plion-Bernier, 1973). No doubt the relative novelty of trains (aptly called in Thai *rot fai* or 'fire vehicle'), heralded by vast billows of smoke, contributed to the general impression of majesty and power associated with the white elephant.

That the Thai people are at once poetic and pragmatic is evidenced in a proverb that is apposite at this point. White elephants (with few exceptions) are invariably born and captured in the jungle. Thus they come from utmost simplicity to utmost regal stature and 'wealth' and all that this implies. The proverb 'Chang pheuak kert nai paa' wisely warns men who have acquired wealth and influence not to forget their roots as 'even the white elephant was born in the jungle'.

In 1855, however, well before the age of steam in Thailand, Sir John Bowring (Fig. 77), though much awaited by King Mongkut, was not able to arrive in time to witness the traditional welcoming ceremonies. Nevertheless, he did eventually visit the king's most recently acquired white elephant, noting the splendour in which the young she-elephant was arrayed, and her frisky lack of appreciation of the 'cloth of gold and ornaments, some of which she tore away, and was chastised for the offence by a blow on the proboscis by one of the keepers'. Bowring was suitably impressed by being greeted by the elephant with 'a salaam by lifting her proboscis . . . more than once at the prince's bidding' (Bowring, 1857: Vol. 2, 312–13).

As a direct result of the cordial relations between the king, and himself as Queen Victoria's envoy, Sir John was to return to England having successfully renegotiated the Treaty of Friendship and Commerce between the two realms. This treaty, which was to pave the way to eventually opening the country to the West and revolutionizing trade in Siam, was sealed with the state seals of the Siamese kingdom (Fig. 78). One of these appositely is 'in the shape of the divine Elephant bearing three heads' called Erawan, the elephant of Indra (Manich Jumsai, 1991: 94).

Entrusted to the care of the returning mission were many rich and valuable state gifts from Siam to the Crown of England. However, on a more personal level, Sir John received some talismanic offerings:

Amidst the most valued presents sent by his Majesty to the Queen Victoria, was a tuft of white elephant's hairs; and of the various marks of kindness I received from the King, I was bound to appreciate most highly

Fig. 77
Sir John Bowring, the British Emissary to Siam in 1855. Engraving from Sir John Bowring, *The Kingdom and People of Siam*, 1857.

a few hairs from the tail, which his Majesty presented to me. (Bowring, 1857: Vol. 1, 475.)

Such hairs, mounted with silver and gold in tassel-like (or now-adays in bracelet) form, were considered, then and now, as most auspicious mementoes. In fact, traditionally they were also incorporated in the ceremonial whisks that form part of the coronation regalia (Quaritch Wales, 1931).

After his departure from Bangkok, Sir John was to receive a most unexpected souvenir of the white elephant that he had recently visited. King Mongkut wrote to him expressing his great regret at the sudden death of 'this curious and beautiful animal' and as a memento sent to Sir John 'a portion of her white skin with beautiful body hairs preserved in spirit.... I trust it will be for an article of curiousity in hand of your Excellency', and signed himself as 'your Excellency's beloved faithful friend ... S. P. P. M. Mongkut, Rex Siamensium' (Historical Commission of the Prime Minister's Secretariat, 1994: 24). In his 1857 report, Sir John mentioned that he subsequently transferred this unique mark of royal favour to the Museum of the Zoological Society of London (Bowring, 1857: Vol. 1, 476).

The Thai custom of preserving part of the skin of deceased white elephants has resulted in a strange and interesting exhibition in the Ivory Room of the National Museum in Bangkok. There, many such samples are preserved in spirits in glass containers, together with a collection of massive ivory tusks of long deceased royal elephants in an eerie almost forest-like display.

Throughout Thai history, the untimely death of a white elephant has been regarded as an occasion of great calamity. In *The English Governess at the Siamese Court* (1870), Anna Leonowens describes a curious and touching incident that took place in 1862. It seems that elaborate preparations for the reception of a newly proclaimed white elephant were suspended upon its sudden death, and the magnificent pavilion, the construction of which King Mongkut had up to that time inspected daily, vanished, at the order of the Kalahom or Prime Minister.

In the evening his Majesty came forth, as usual, to exult in the glorious work. What was his astonishment to find no vestige of the splendid structure that had been so nearly finished the night before. He turned, bewildered, to his courtiers, to demand an explanation when suddenly the terrible truth flashed into his mind. (Leonowens, 1870: 144–5.)

Unlike the seventeenth-century monarch who condemned to death those deemed at fault, King Mongkut, no tyrant, though having the traditional powers of Chao Chivit, or Lord of Life, confined himself to tearful melancholy.

Siamese admiration and respect for the white elephant were sometimes to be expressed in ways unexpected in the West. After the departure of Bowring, a mission of Siamese Ambassadors, bearing letters of friendship, credentials, and gifts to Queen Victoria, was dispatched to England in late 1857. Despite

Fig. 78
A line drawing of the Phra Maha Eyerabote [Erawan] Seal used on State documents in the nineteenth century. From Seni Pramoj and Kukrit Pramoj, *A King of Siam Speaks*, 1987.

understandable difficulties in adjusting to the British winter, the mission was most fruitful (Manich Jumsai, 1991). On their return, the report of the Ambassadors provided Siam with its first window on the West since the seventeenth century. Said to be included in the report was a remarkable observation of the personal appearance of Queen Victoria:

One cannot but be struck with the aspect of the august Queen of England, or fail to observe that she must be from a race of goodly and warlike kings and rulers of the earth, in that her eyes, complexion, and above all her bearing, are those of a beautiful and majestic white elephant (Leonowens, 1870: 145).[4]

King Mongkut's successor, his son Chulalongkorn, who became king in 1868 at the age of fifteen, was to bring closer to fruition the modernization of Siam, initiated in the previous reign. However, while his numerous reforms radically changed the face of Siam, ancient ceremonial and traditions were not forgotten. In keeping with such traditions, in order to reward distinguished service to the Crown, King Chulalongkorn enlarged the scope of the Most Exalted Order of the White Elephant, an Order originally initiated in 1861 by King Mongkut. Greatly prized, the Order's insignia is a star-like emblem, the centre of which features a white elephant on a ground of dark red enamel. Appropriately coloured ribands signify the rank of the award. Of all the royal orders, it is only this Order that is also conferred by the Crown on foreigners. The first recipient, in fact, was Queen Victoria. In the twentieth century, the Order has been bestowed on distinguished foreign scholars, heads of businesses that have contributed greatly to Thailand's economic development, foreign envoys, and visiting heads of state. It is to be regretted that the Western press largely prefers to ignore this as a mark of singular honour and delights in recalling the erroneous English meaning of the term, and thus mocking the recipient to score political points.

In 1882, at the time of the Centenary of Bangkok, King Chulalongkorn also dedicated monuments to his vigorous and visionary ancestors, the kings of the Chakri Dynasty. These three memorials are set around the Royal Pantheon in the grounds of the Temple of the Emerald Buddha and take the form of tall marble plinths surmounted by decorative emblems of the respective kings. Each monument is surrounded by bronze replicas of the white and other important elephants of the respective kings' reigns, thereby reaffirming the significance of the white elephant as a traditional symbol of religion and kingship in Siam at that time (see Fig. 19).

---

[4]Ironically, it is in Bangkok, never part of the British Empire, that until relatively recently homage was still paid regularly to Queen Victoria in the form of flower garlands, incense, and offerings being placed at the massive and rather forbidding bronze statue of the Queen in the grounds of the British Embassy, when they were still open to the public in pre-terrorist days. It seems that this 'shrine' enjoyed quite a reputation for granting fertility to supplicants.

In like manner, during the restoration of the Temple of the Emerald Buddha for the Bicentennial of Bangkok, celebrated in 1982, a fourth and similar monument was erected and dedicated retrospectively to Rama VI, Rama VII, Rama VIII, and the present monarch, King Bhumiphol Adulyadej or Rama IX. It is this latter monument that is perhaps the most well-known, as tourists, encouraged by their Thai guides, constantly pose beside it for photographs and keep the foreheads of some of the favourite bronze elephants permanently shiny by countless pats for good luck (Fig. 79). Unlike any of the other decorative and often fantastic statuary in the temple complex, all the memorial elephants are realistically wrought. At the same time, they are a permanent

Fig. 79
Dedicated during the Bangkok Bicentennial in 1982, this monument, with replicas of royal white elephants, honours Rama VI, VII, VIII, and IX, Temple of the Emerald Buddha, Bangkok. (Photograph Rita Ringis)

111

reminder of the respect in which the legendary white elephants of Siam have been held for centuries.

In 1932, Siam became a constitutional monarchy, and since 1939 it has been known officially as Thailand. Since that time, as the kings were no longer considered Chao Chivit or Lords of Life and all that this title of absolute monarch implied, it would seem likely that respect for white elephants as symbols of divinely ordained monarchic power might be somewhat outmoded. In fact, these symbols still have relevance, as the strength of Thai tradition is such that while His Majesty the King is above politics, his moral power and example are manifest in all aspects of Thai public life. All major ceremonial and festivals, whether civil or religious, continue to be presided over by the monarch or his representatives.

Today, the concept of the 'white elephant' still grips the Thai imagination somewhat. However, in keeping with modern times, reception and care of these 'divine creatures', while retaining traditional respect, no longer involves the sumptuous ceremonial as described by observers in the past. In former times, the white elephants were stabled in the grounds of the Grand Palace, adjacent to the Temple of the Emerald Buddha, the founding area of Bangkok. There, in the nineteenth century, European travellers with the right introductions were able to visit the white elephants and record their at times less than favourable impressions for posterity.

The Grand Palace is now reserved for ceremonial use, and the only elephants resident there are the monumental bronze sentinels at the foot of the staircase to the Chakri Maha Prasat Throne Hall (see Fig. 20). Many of the other palace buildings are now the offices of the Bureau of the Royal Household, through whose gracious assistance this twentieth-century observer was privileged to receive royal permission to visit the white elephants of the present reign, housed elsewhere.

In the centre of Bangkok is an island-like estate where the royal family resides. Known as Chitralada Villa or Palace, it is surrounded by a lotus filled 'moat' or canal. Beyond that, enclosed behind an encircling wall and prolific greenery lies a veritable mini-city, to which access is through gates manned by security forces. A journey through the outer areas takes the visitor past low-rise office buildings and the headquarters of the famous Chitralada Shop, where traditional village crafts revived from possible extinction are displayed and sold for the benefit of rural people by Her Majesty's Foundation for the Promotion of Supplementary Occupations and Related Techniques, popularly known as SUPPORT.

Further tree-lined streets lead past a vast garden-like setting at the centre of which is an extensive villa which flies the standard of His Majesty, when he is in residence: the red Garuda on a yellow background. Beyond that, along a winding road, here and there amidst the foliage, are visible occasional modest buildings indicating perhaps a laboratory, perhaps a model farm. Data and results from scientific and agricultural experiments conducted here are integrated and applied in the many ongoing projects initiated by His Majesty for the welfare of the rural people of Thailand.

The road comes to an end at what appears to be a small forest, in the midst of which is visible a complex of Thai-style roofs (Fig. 80). Surrounded by greenery, as well as a modified version of the traditional elephant kraal or wall of retaining pillars, these are the Royal Elephant Stables. Here reside in ordered and neat simplicity the famous but rarely seen royal elephants, as well as their keepers, four per elephant, in simple traditional quarters adjacent to those of their charges. The atmosphere is unhurried and rustic, and it is only the occasional siren from the traffic humming beyond the greenery that makes one aware that this is in the heart of Bangkok.

The elephant keepers, though dressed in the familiar khaki everyday uniforms of all civil servants, are clearly up-country folk. In charge of them is their Nai Dee, or 'good master', a respected mahout in his forties, delighted to talk about his charges to this willing listener. He notes that the royal elephants, as befits their status, have a refined palate. Their taste in grass demands fresh and succulent samples which he and his colleagues seek out every three or four days, travelling out with their trucks well beyond the reaches of unfortunately polluted Bangkok. White or auspicious elephants, he affirms, do not drink from common water areas. In fact this refusal to join communal habits which are customary in

Fig. 80
Traditional-style pavilion of a white elephant at the Royal Elephant Stables, in the grounds of Chitralada Palace. (Photograph Rita Ringis)

113

elephants is a sign of the uniqueness of a particular elephant. He notes also, confirming the bemused comments of foreign observers of the distant past, that these auspicious elephants have a distinct sense of their own dignity, and grow melancholy if poorly treated.

The chief mahout's gap-toothed smile is in constant evidence as he lovingly describes the virtues of 'his' elephants. Like the rest of the keepers, he hails from the provinces, and started his apprenticeship as a mahout when still a youngster, under the guidance of his father, who came from a long line of 'elephant men'. Despite the longevity of the family tradition, Nai Dee confesses, with a mixture of regret and pride, that his sons will not be continuing the profession, but are studying instead at college in Bangkok.

Supervising the activities of the Stables, and the health of the elephants is a young veterinarian, Dr ML Phiphatanachatr Diskul. In the past, because of the importance of the royal elephants to the safety of the realm, those in charge of them traditionally came from the nobility. Dr Phiphatanachatr's title of Mom Luang (ML) indicates that this tradition is still upheld. As a scientist, Dr Phiphatanachatr makes use of the latest techniques to ensure the well-being of his charges. Yet he is also very much aware of the usefulness, and he notes, the reliability of information available in the ancient Thai treatises on distinguishing characteristics, for good or ill, of elephants. These writings in their beautifully illustrated manuscript form may appear superficially to be merely poetic and thus thought by some as superstitious and quaint. They are, however, based on centuries of traditional observation and intimate knowledge and understanding of elephants. Dr Phiphatanachatr affirms that modern scientific observations about elephants frequently confirm the validity of the keepers' folk wisdom, much of which is detailed in the manuscript literature (see Figs. 3, 9, and 37).

Both the scientist and the keepers defer to a frail, ascetic-looking, and much respected man, Khun Svest Dhanapradith, who has long served the Bureau of the Royal Household. Born well over eighty years ago, Khun Svest started as a humble clerk in the Royal Household, but was for many decades in charge of the rites of confirmation of the authenticity of a white elephant, and the ceremonial associated with its formal induction to the equivalent in rank of a prince. Though now officially long retired, of his own choosing Khun Svest devotedly attends the Royal Household daily as an adviser. As a walking encyclopaedia of oral traditions, his wealth of knowledge in matters of pachyderm protocol is considered invaluable and he is much sought after in this regard.

In fact, for that purpose, prior to actually visiting the Royal Stables at Chitralada, this twentieth-century observer had met with Khun Svest in one of the many handsome nineteenth-century buildings now occupied as offices by the Bureau of the Royal Household in the Grand Palace complex. Khun Svest frequently referred to his *aide-mémoire*, a large photograph album of candidates and ceremonies celebrating previous investitures of white elephants, and explained that while the general disposition and character of each elephant candidate were observed in detail, sometimes over a period of many months, seven main characteristics were essential before an elephant could be considered as 'white' or *chang pheuak*.

Directing the interview, from the Office of His Majesty's Principal Private Secretary, was Mom Rachawong Putrie Viravaidya whose lively prompting and gentle guiding of Khun Svest's encyclopaedic and wide-ranging memory ensured that the meeting was both stimulating and informative. Switching rapidly from Thai to English, Khunying Putrie, as she is popularly known, kindly provided the essential bridge whenever this observer's understanding wavered at the effort of keeping up with unfamiliar terminology.

By tradition, according to Khun Svest, the conditions for presentation of an elephant as 'white' are exclusive and restrictive for obvious reasons: the value of a symbol of royal virtue would be considerably lessened by excessive availability. Within those restrictions, however, an elephant candidate that closely approaches but does not fulfil all of the desired characteristics is none the less out of the ordinary, and therefore is likely to be accepted as an auspicious elephant, *chang samkhan*. This category may be further subdivided into unusual coloured elephants, *chang si pralaat* and *chang niem*. However, for an elephant to be deemed pre-eminent and white, *chang pheuak*, the conditions are most rigorous.

First, the eyes must be white, with a pale iris of lustrous sheen, and bulge in a manner reminiscent of the eyes of Siva's legendary bull, Nandin. Secondly, the palate should be of a white or pale unspotted colour, similar to that of a lotus. Thirdly, the hoof nails should have a shell shape, and be pink and smooth, no doubt a sign of good circulation and therefore of the health of the candidate. When it is remembered that the untimely death of a white elephant is considered as an omen of ill fortune for the country, such exclusive characteristics are understandable. The most pre-eminent elephants would have five nails on each hoof, a most rare occurrence.

Fourthly, the colour of individual bristles or body hairs of the elephant are also taken into detailed account. Most unusual and therefore most auspicious are two hairs growing out of one follicle. These hairs must not only be white, but even the tip of the hair in the follicle must be white. Experience has shown that if the colour of the root is dull, the hair will eventually turn black, as will the elephant. (This is not ultimately a bad thing as a black elephant is also classified as auspicious, but of a different class.) Not only is the white colour of the hairs important, but their distribution is indicative of the white elephant's unusual nature. On its spine, elegantly bent like a bow, the hairs should be stiff and upstanding, not unlike that of a ridge-back. However, at the back of the thighs, the hair must be sparse, while at the ears, it must be prolific: in fact, the bushier the better, rather like the sideburns or mutton-chop whiskers of former European male fashions. The desirability of this pale hairy quality is emphasized by the custom of decorating state elephants with ceremonial yak hair whisks (to increase the hairiness as it were) worn at the ears, which by tradition must have long, smooth, and supple lobes like those of the Buddha. Overall ear size also matters: the ears should be of sufficient magnitude to allow both ears to meet, if folded, on the forehead of the elephant.

Fifthly, the overall body skin must be of a pale colour, rarely actually white but more likely of a pale terracotta hue as against the usual grey tones of the ordinary elephant. Sixthly, the tail, unlike the stumpy brush of the everyday elephant, must be long and straight, and the hairs at its tip must be luxuriant and white, forming a halo of bodhi-leaf shape. Seventhly and finally, the genital area must also be of a white or pale rufous colour, like that of the sheath-like petals of the dried flower of the banana tree. Even the customary manner in which the candidate elephant relieves itself is taken into consideration, as in all things great or small, it must exhibit a genteel noble mien.

While these seven areas are considered essential, other aspects are taken into lengthy consideration. For example, if the elephant is a tusker, its tusks must curve in harmony with the trunk. Asymmetrical or crossed tusks, frequent in everyday elephants, are rarely acceptable. However, if the tusks do cross, only the right tusk crossing towards the left is appropriate. In fact, the varying combinations of all the unusual as well as usual characteristics of the elephant also determine the 'family' or caste into which the elephant is further categorized. Taken into consideration here, for example, is the relative size of the head, the eyes, the chin, and the feet as well as the quality of hair and skin texture. (Outlined in Chapter 3 is the myth of the origins of the four families of elephants, Phromapong, Issuanapong, Vissanupong, and Akkanipong, created from the petals and stamens of the lotus of the Hindu god Vishnu.)

Of great importance also is the unusual quality of the sound or call produced by the elephant in question. According to Khun Svest, that of a white elephant has a low, melodious plangency, similar to the plaintive tone produced by gently breathing into the conch shell, the instrument Phra Narai (the Hindu god Vishnu) used to awaken primal sound during the creation of the universe. The conch shell, with all its associations with the powers of divinities, is still used today in Thai traditional ceremonial.

During the interview with Khun Svest, lively attempts to match the exact quality of this desirably auspicious elephant sound lead to something of a cacophany, startling attendants day-dreaming at the sunny windows of this vast room, outside of which are visible the dozens of daily tourist visitors to the Grand Palace marvelling at its glittering splendour in the mid-morning sunlight. But here, inside the cool dim room, grandeur and pomp are moment-

116

arily forgotten as Khun Svest's great age and experience prompt him to gently but firmly reject as totally inept Khunying Putrie's extremely helpful attempt to mimic the sound of a white elephant calling. The esteemed master's own sonorous variation, produced while he cups his ear the better to fully experience the plaint's wavering and apparently auspiciously elephantine timbre, is clearly a far more acceptable approximation. Both performances are greatly appreciated by the twentieth-century observer whose own attempts do not even merit serious consideration.

Theory is put to practice on the day of the actual visit to the Royal Elephant Stables at Chitralada Palace. However, that special, distinguishing white-elephant call, so ineptly rehearsed at the Grand Palace interview, is now considerably more resonant, being genuine, and cajoled from a decidedly real white elephant by Khun Svest, assisted by the blandishments of ever-helpful stablemen and keepers. The trumpeting elephant, Phra Savet Adulyadejbahon (Fig. 81), very parchment-pale and almost amber-eyed, sporting enormously long and superbly symmetrical upcurving tusks, resides in a traditionally cruciform building which is his 'palace', the other elephants being housed in more functional recent structures.

Unlike the pavilions with gold and silver floors reported by Neale in the nineteenth century, Phra Savet Adulyadejbahon's residence is simple, clean, and austere. The mahouts and attendants of this princely elephant, proud of his cleverness, tease him into

Fig. 81
Phra Savet Adulyadejbahon, a white elephant of princely rank, at the Royal Elephant Stables. (Photograph Rita Ringis)

117

good humour, clucking at him shamelessly, until he produces a suitable salaam, obediently kneeling his great bulk and curving up his trunk in greeting for the astonished and delighted visitor.

While a 65-year-old female elephant, Pung Pun, is the oldest at the stables, Phra Savet Adulyadejbahon, at forty-two is the oldest male white elephant of the present reign. Indisputable doyen of the stables, he amiably salutes once more, trumpeting his comment on the whole process, and then the 'audience' is at an end. Wearied by his exertions at holding court, Phra Savet Adulyadejbahon retreats into a somnolent dignity, and it is time to visit the other ten royal elephants at the stables.

Twenty elephants considered auspiciously significant have been presented to the Crown since His Majesty King Bhumiphol Adulyadej acceded to the throne in 1946. Of these twenty, twelve are extant but only eleven reside at the Royal Stables, the twelfth elephant residing still with its owner, presumably awaiting an auspicious time for formal presentation to the Crown. This appears to be likely sometime between mid-1995 to mid-1996 when the Thai nation will celebrate the Golden Jubilee of the Accession to the Throne of His Majesty King Bhumiphol Adulyadej, the longest reigning monarch in the world (Fig. 82).

Fig. 82
Emblem designed for the fiftieth anniversary celebrations of His Majesty King Bhumiphol Adulyadej's accession to the throne. (Courtesy of the Tourism Authority of Thailand)

118

At present at the Royal Stables, six of the elephants (three males and three females) fulfil the rigorous demands required of the *chang pheuak* or white elephant. Classified as members of the Phromapong family of elephants (the family of Brahma) and the Vissanupong family (the family of Vishnu), these elephants have been inducted into noble rank. The other five resident elephants who do not quite fully conform to the necessary characteristics are nevertheless also in the care of the stables, being deemed *chang samkhan*, or auspiciously significant elephants (Fig. 83).

Apart from the elderly Pung Pun and Phra Savet Adulyadejbahon, who is in his middle-aged prime, the other elephants are mere youngsters, some barely out of their teens. In keeping with modern trends, these elephants are able to lead lives less circumscribed by tradition, and in 1987 three of them spent time upcountry as students in the care of the Elephant Training School near Lampang. Clearly, the benefits of such 'vacations' are so obvious that another Royal Elephant Stable is projected for completion near Lampang in the near future. There, from October to February each year, both the white and the auspicious elephants will have the opportunity 'to live freely in the jungle' all day, spending their nights in the security of the stables (Phiphatanachatr Diskul, pers. com.). However, at the stables at Chitralada Palace, all is not restrictive and stuffy, for the elephant residents enjoy a varied social life that includes frequent informal visits from members of the royal family.

Fig. 83
Auspiciously significant elephants (*chang samkhan*) strolling in the grounds of Chitralada Palace. (Photograph Rita Ringis)

Meanwhile today, on this observer's visit, the decidedly frisky youngsters assemble outside their stalls for exercise and play. Their colours vary from almost black (which actually comes as quite a shock) with pale speckling and stippling, similar to that seen in traditional paintings of elephants, to a warm mud-coloured chalky yellow. In a slow and stately procession, they wend their way beyond the stables towards the lake in the Chitralada grounds, past a little house near the stables in which resides a lively white monkey, chattering at passers-by. Tradition holds that a white crow, a white monkey, or both, should reside near the home of the white elephant to protect it from evil and misfortune. In this case, these auspicious elephants, accepting the monkey's pale propitiatory presence, pay him no heed, and with lithe and graceful gait continue on their way.

At the lake, auspicious status is forgotten as the elephants in all their massive bulk plunge and splash monumentally into the water, frolic about, and thoroughly enjoy themselves like the teenagers that they are (see Fig. 46). Much water is displaced and sprayed, and after bathing, the elephants visit adjacent trees for a vigorous scrub and rub before the whole group, with adolescent reluctance, obeys the orders of the mahouts, and sets out once again in a dignified manner back to the stables. There, a further effervescent wash and rinse, helped by the diligent mahouts, takes place at a little canal built across the lawn of the stables. After much obviously amiable and affectionate elephantine play (which includes the placing of their vulnerable trunks into each others' mouths), the elephants return sedately to their individual stalls for 'quiet time', and an air of rustic peace descends.

Here, in the forested grounds of Chitralada Palace, surrounded by the sprawling and rapid development of Bangkok as a late twentieth-century boom town, these auspiciously significant elephants abide as symbols of ancient traditions and noble virtues.

6 Elephants in Thai Art and Mythology

Fig. 84
Elephant and rider design in
supplementary weft on plain warp and
weft of handspun cotton textile.
(Photograph courtesy of Tilleke and
Gibbins Collection of Southeast Asian
Textiles)

Fig. 84
Elephant and rider design in supplementary weft on plain warp and weft of handspun cotton textile. (Photograph courtesy of Tilleke and Gibbins Collection of Southeast Asian Textiles)

GIVEN the ubiquity of the elephant in almost every aspect of Thai life of the past, it is little wonder that its image has been consistently captured in the arts and crafts of Thailand. The elephant's form has been depicted over the centuries in a wide variety of media, ranging from now-weathered sculptures in stone or stucco adorning ancient temples to crisply geometric silhouettes woven into the colourful silks and cottons of the present (Plate 20; Fig. 84). The vitality and diversity of the numerous portraits of the elephant reflect not only the Thai people's affinity with this engaging animal, but also their recognition of its role in the history and traditions of the land. After all, although the water-buffalo is also ubiquitous in the life of the Thai people, it is rarely eulogized in art.

Many examples of lively depictions of the elephant represented in a variety of art and craft forms are currently held in museums and private collections. In fact, it is something of a miracle that many such delicate pieces have withstood the buffeting of time comparatively better than have the grand temples of former ages. For example, dating from some seven hundred years ago are small ceramic votive figurines, produced for the local market by the thriving export ceramics industry around Sukhothai and Si Satchanalai of the time. Enhanced by the famous jade-like celadon glaze used also on the fine export ceramics, these elephant replicas, some bearing riders and howdah-like containers clearly intended as oil-lamps, were destined no doubt for domestic propitiatory shrines (Fig. 85). Such simple, naïve, and homely objects remind us of the daily hopes and fears of ordinary people of the great kingdom of Sukhothai, renowned for its majestic temples and vast Buddha images.

Possibly the most famous elephant in the art of Thailand was that recovered intact from the crypt of Wat Rajburana in Ayutthaya by the Fine Arts Department during restoration work in the 1950s (see Chapter 4). Portrayed as kneeling in homage to the Buddha, the little elephant figurine, crowned by an intricate diadem, holds a finely worked bejewelled garland in its upfurled trunk. Two short ornamented protuberances below the trunk suggest that ivory may have originally been sheathed in them as its tusks. On its back is a minutely worked part-filigree howdah, secured by golden chains and jewel-encrusted trappings. Similarly

121

Fig. 85
Celadon-glazed oil-lamp of elephant and
rider, 22 cm high, fourteenth–fifteenth
century, National Museum, Bangkok.
(Photograph courtesy of the National
Museum Volunteers, Bangkok)

worked anklets encircle the elephant's legs. As befits such a noble
elephant, its tail is set in a graceful curve and terminates with an
appropriate bodhi-leaf shaped filigree halo (see Plate 19).

Many other fragile objects on which the image of the elephant is
portrayed have survived the test not only of time but of a climate
that hastens decay. This may be due partially to the fact that many
of these objects were created as expressions of religious beliefs,
and as such were treated with the care those beliefs inspired.
Simple articles of everyday use such as cabinets or even food containers, suitably embellished by the craftsman's skill, were transformed into vessels fit for ritual and ceremony only.

Salvaged in recent times from benign neglect in up-country
temple locations, and displayed today in the Bangkok National
Museum, as well as in the National Library of Thailand, are fine
collections of wooden scripture cabinets covered in delicate black-
and-gold lacquerware designs (*lai rod nam*) depicting lively animal
and human figures set amidst realistic or fantastic flame-like
(*kranok*) vegetation (Plate 26). In these mythical forests, elephants
are sometimes depicted here and there either in a realistic manner
or in a richly stylized way as vehicles or mounts of protective
Hindu gods and heroes of Buddhist folk-tales. Such black-and-
gold lacquerware designs also decorate door and window panels
of temples of the past and present day.

122

Similar traditional motifs were also produced by the mother-of-pearl inlay process (*kruang muhk*) on lacquered screens, doors, furniture, medicine boxes, musical instruments, and ceremonial vessels. Set into the sombre black lacquer background, the intricate patterns of thousands of pieces of the milky mother-of-pearl shell scintillate like fire opals. Door panels at the Temple of the Emerald Buddha and at Wat Rajabophit are fine examples of this meticulous craft. At the latter, apposite to elephants, door panels display various royal orders rendered in mother-of-pearl, including that of the Most Exalted Order of the White Elephant for distinguished service to the Crown.

Some of the finest examples of this craft are displayed in the Gallery of Mother-of-Pearl objects in the Bangkok National Museum. These include a scripture cabinet whose door panels depict the three-headed Erawan elephant ridden by the god Indra, protecting the Buddhist scriptures originally contained within (Fig. 86). Particularly notable in the display is a highly decorative cover of a monk's food bowl used during celebrations for the

Fig. 86
Indra on the three-headed Erawan elephant. Detail of a cabinet with mother-of-pearl inlay, National Museum, Bangkok. (Photograph Rita Ringis)

123

coronation in 1873 of King Chulalongkorn (Rama V). Set at the centre of the cover is the three-headed Erawan elephant, symbol of a king's virtue. The inlayer's extraordinary skill is evident in that from whatever angle one looks at this mythical elephant, three heads would appear to be the norm in elephants. Further enhancing that illusion is the artist's judicious choice of the particular pieces of mother-of-pearl that cause this illustrious white elephant to glow with a superb silky pink and green lustre.

Fittingly, by tradition, the elephant was rarely represented in ivory. Generally speaking, when ivory was used by the Thai as a medium, it was reserved, in the main, for objects of royal ritual or religious use. Thus finely carved ivory panels formed the ornate covers between which traditional palm leaf manuscripts of royal editions of the scriptures were placed; ivory also enhanced musical instruments, and the handles of ceremonial swords and knives, their glossy paleness somewhat mellowed by climate and handling.

Displayed in the Suan Pakkad Palace Museum and the Bangkok National Museum are numerous small skilfully carved ivory objects as well as the actual tusks, some of enormous size, of royal elephants of the past. That the tusks have been kept in their original state, as removed from the deceased elephant, and not cut up and carved into ivory objects attests to the respect in which the original elephants were held. These tusks were considered literally as relics of the past. Contrasting with pristine tusks in the collections are several pairs still retaining the tusk shape, but into the surfaces of which have been carved meditating Buddha images set in ornamental niches. Some also display the carver's exceptional skill in the filigree almost mesh-like quality of their carved ivory surfaces below which further carvings are visible.

Extraordinarily fine workmanship is also evident in a unique ivory howdah (*kub chang*) on display in the Transportation Gallery at the Bangkok National Museum (Fig. 87). Originally presented to King Chulalongkorn by the Prince of Chiang Mai in

Fig. 87
Ivory howdah presented as tribute from Chiang Mai to King Chulalongkorn in the early twentieth century, now at the National Museum, Bangkok.
(Photograph Rita Ringis)

the early twentieth century, this delicate howdah, constructed of panels of ivory carved in intricate filigree, belies its essential sturdiness. It is, after all, essentially a chair, to be placed on an elephant's back. Howdahs were usually made of wood or rattan, and if intended for ceremonial use, they were carved and gilded. Today in Thailand, replicas of howdahs are popularly used in interior decoration as lounge chairs.

In fact, continuing traditions of Thai craftsmanship allow even the humblest to acquire some object associated with the elephant, often based on fine art works of the past. Thus the ancient art of bronze casting, usually reserved for Buddha images, also produces a vast variety of figurines, including elephants both real and mythical. Nowadays, on sale at amulet markets, temple fairs, and even department stores are small replicas of the Erawan elephant ridden by Indra. Bronze statuettes and tiny replicas of the elephant-headed god, Ganesha, are also sought after, by students in particular, to ensure success in scholastic ventures (see Fig. 97). Popular too are small kneeling elephant figurines bearing a lotus in the trunk, reminiscent of the important role of the elephant in the life of the Buddha (Fig. 88). Any such images may be purchased, and when blessed by the monks, kept in places of honour in the home. Such figurines are, in fact, metaphors for traditional beliefs, for they immediately call to the devotee's mind the roles of the elephant whether in the life of the Buddha, or in the ideal worlds of the gods. In fact, the arts and crafts of Thailand are not confined to museums and galleries where the élite may congregate, but are accessible to all, in small personal propitiatory effigies, or on a grand scale, in the temples that abound in the land.

To fully appreciate the variety of Thai impressions and depictions of the elephant in all its forms, it is useful to understand its background in the larger context of Hindu and Buddhist mythology. While the elephant's various roles are briefly listed in numerous dictionaries of mythology, this chapter, rather than merely

Fig. 88
Kneeling bronze elephant figurine holding a lotus, 9 cm high, commonly available in amulet markets and department stores. (Photograph Rita Ringis)

cataloguing those myths, is concerned with illustrating them by describing traditions of that mythology in Thailand as manifested not only in folk beliefs, and customs of the past and present, but also in the traditional arts.

Until very recent times, the arts of Thailand have been devoted entirely to the expression of religious themes. In keeping with this tradition, the artists and craftsmen have been for the most part anonymous, their aim being the portrayal of a shared religious perspective, not of a unique personal vision. Thus the architecture, sculpture, and painting, and even the so-called 'minor' or decorative arts of Thailand, were in fact to a greater or lesser degree visible expressions, in universally recognizable forms, of Hindu and Buddhist concepts adapted and developed over the centuries into a distinctly Thai idiom.

For many centuries now, the Thai have followed the tenets of the relatively austere Theravada Buddhism in which the supernatural has no place, the Buddha being regarded not as a god but as the Great Teacher whose doctrines and example each individual may follow on the road to enlightenment. However, earlier Hindu elements of divinities and their attendants have not been abandoned but incorporated into temple decoration and design, to provide a framework for the universe and its mysteries.

This has led to an apparent paradox: Thai temples, while enshrining Buddha images of graceful simplicity and spirituality, are glitteringly palatial in appearance. Their sumptuous decoration traditionally features elements of lively fantasy, of worlds clearly other than this. Window panels, doors, and pediments of assembly halls abound with depictions of luxuriant flame-like foliage rarely seen in nature. Many armed deities and strange composite animals, including three-headed elephants, emerge from these fantastic forest decorations (see Plates 9 and 10).

Of this apparent fantasy world, in what appears to be one of the earliest critiques of Thai artistic traditions, in the seventeenth century a French observer noted of the Thai that

their fancy is to slight and disesteem whatever is after Nature only. To them it seems an exact Imitation is too easie, wherefore they overdo everything. They will therefore have Extravagances in Painting as we will have Wonders in Poetry. They represent Trees, Flowers, Birds, and other Animals which never were. They sometimes give unto Men impossible proportions, and the Secret is, to give to all these things a Facility, which may make them to appear Natural. (La Loubère, 1693: 71.)

This observation-cum-backhanded compliment is in fact absolutely correct, and while then intended as criticism could today be perceived as praise. To the sympathetic observer, these 'Extravagances', 'Animals which never were', and 'Men [of] impossible proportions' are not merely arbitrary decorations that, as it were, 'overdo everything', but in fact form an integral part of the symbolism of Thai monastic structures. Indeed, the Thai temple represents an earthly embodiment of a metaphysical universe inhabited by gods and other celestials. In its layout and decoration, the temple serves as a complex metaphor to the faithful, alluding to the path that the devotee must ascend on his spiritual journey towards liberation and enlightenment.

Emphasizing the temple's otherworldly and symbolic aspect is the wall that separates it from the 'real' and everyday world (see Fig. 69). Within the wall's enclosing boundaries lie the various structures common to all Thai temples, including the rectangular ordination and assembly halls (*ubosot* and *viharn* respectively); the sacred relic chambers, whether in the bell-shaped *chedi* form, generally of smooth or simple surface decoration, or in the form of tall slim towers, the *prang*, generally of highly redented or ornamented surfaces. The most famous examples of these are the vast golden *chedi* at the Temple of the Emerald Buddha, and the soaring *prang* tower at Wat Arun, the Temple of the Dawn.

Each of these types of structures, while they are present-day symbols of Thai Theravada Buddhism, represents separate traditions that have been adapted by the Thai: the *chedi* is essentially an originally Buddhist structure, the *prang* derives from originally Hindu concepts. Both structures, in their form and symbolism, embody concepts of ascent towards higher levels of being. Both structures have also associated with them, not at all frivolously, elephants as symbols of worlds other than the mundane.

However, while Hindu and Buddhist traditions have their roots in India, many of the concepts as embodied in Thai temple architecture and decoration did not come directly from India but were filtered over time through other regional cultures. Thus on the one hand, Thai contacts with Sri Lankan Theravada Buddhist traditions after the thirteenth century led to the introduction of the bell-shaped *chedi* form adapted from the bulbous dome-shaped Sri Lankan *dagoba* or sacred relic chamber, itself originally derived from Indian relic mounds or stupa prototypes.

On the other hand, the gradual evolution in form and symbolism of the Thai *prang* tower, in addition to general principles of temple layout, owes much to influences absorbed from the Khmer civilization. Between the seventh and mid-thirteenth century, Khmer political power in regions of present-day Thailand was reflected in the construction on pilgrimage and trade routes of many stone and laterite sanctuaries, at first sacred to Hinduism and later to Buddhism. Lying in scattered regions of Thailand, these sanctuaries, many recently restored, form an integral part of the Thai cultural heritage.[1]

The Thai transformed these essentially earthbound Sri Lankan

---

[1]Scholars comparing the sanctuaries constructed in Cambodia itself and those built in the Thai regions until the thirteenth century have noted consistent subtle stylistic differences between them indicating that the Khmer structures in Thailand are not mere copies of the metropolitan style of Angkor. To distinguish the two, therefore, Khmer Art in Thailand is also sometimes called Lopburi Art after the name of a Khmer stronghold there in the twelfth century.

and Khmer prototypes over the centuries, elevating and elongating them into the now familiar lofty, slender, and distinctly Thai monuments that draw the eye and spirit upward. In both cases of cultural borrowing, however, the Thai retained the original symbolism of the adapted structures, incorporating it into their own later temple design and decoration.

Like its later Thai temple counterpart, the entire Khmer sanctuary complex, with its outer walls enclosing processional galleries and inner religious structures, was intended as a replica on earth of the universe, an architectural map of the macrocosm as it were (Fig. 89). The most sacred area of the complex was a central sanctuary building topped with a corn-cob shaped tower or *prang*. This stone man-made tower represented the axis or metaphysical centre of the Hindu and Buddhist universe, Mount (Su)Meru, on whose Elysian slopes, according to traditional beliefs, dwelt various powerful deities, lesser celestials, as well as mythical animals, including families of celestial elephants.

Accordingly, as befitted an earthly replica, the outer surface of the central tower or man-made Meru mountain was lavishly decorated with carvings of stone foliage and a rich diversity of celestials (Fig. 90). Sculptural narratives of the exploits of powerful gods as well as lesser beings decorated the lintels and pediments above the entrance doorways to the inner sanctuary, the *garbagriha* or 'womb'. Enshrined within this interior chamber was

Fig. 89
Aerial view of Prasat Khao Phnom Rung, Buriram Province, tenth–thirteenth century. Enclosed by the remains of stone galleries, the east-facing extended cruciform antechamber provides the main entrance to the sacred inner chamber of the sanctuary. (Photograph Banchob Maitreejit, courtesy of the National Archives, Bangkok)

Fig. 90
The southern face of the main sanctuary at Prasat Khao Phnom Rung, Buriram Province, tenth–thirteenth century. Processions with elephants appear on the pediments of the upper body of the *prang*, as well as on the lintel above the corridor entrance to the right. (Photograph Rita Ringis)

the stone *linga* or phallic icon, representing the potency of the god Siva, if the sanctuary was Hindu. If the shrine were Buddhist, an image of the Buddha presided there.

Esoteric and élite ceremonies were held there invoking divine aid to ensure fertility within the land. In Khmer beliefs, the earthly ruler was identified with the god thus invoked and consequently was venerated as *devaraja* or god-king.[2] The act of building a replica on earth of the magic mountain confirmed a ruler's link with the divine. In fact, these stone sanctuaries could be said to have had both religious and political significance. The Thai term

---

[2]Aspects of this cult were adapted by the Thai, particularly in seventeenth-century Ayutthaya. Echoes of this practice still resonate today and are especially evident in royal funeral rites.

129

for this type of sanctuary complex is *prasat hin* or 'stone palace', encapsulating its function: a palace on earth fit for the mystical presence of the gods of Mount Meru.[3]

Replicating on earth the home of the gods ensured their protective and beneficent presence. In effect, the entire sanctuary was an invocation in stone for abundance, prosperity, and potency on earth as bestowed only by the gods through the intervention and participation of the ruler in appropriate ritual. The numerous carvings on these sanctuary towers can be 'read' as texts or books in stone of the *Mahabharata* and *Ramayana* epics—lively local interpretations of Hindu mythology. Any sympathetic visitor familiar with the iconography or vocabulary of the Hindu epics can generally grasp key sculptural scenes or nuances, and thus decipher the significance of the decorative lintels, pediments, and even peripheral carvings at the various levels of the magic stone mountain.

If today's visitors to the Khmer sites in Thailand were to concentrate merely on representations of elephants (Spot the elephant!), they would not have far to look. Catching the viewers' eyes here and there may be fragments of battle or procession scenes featuring caparisoned realistic elephants, with their stalwart human attendants, bringing to life episodes from the *Ramayana* and *Mahabharata* epics, but also providing a glimpse of the human and social context of former times, revealing in vignette form the disposition of troops, as well as war elephants and their trappings (Figs. 91 and 92).

Depending on the geographic region and age of the temple, certain elephant motifs would become evident, as they frequently recur for symbolic purposes on lintels, pediments, and antefix decorations. Many of these representations are no longer at the actual sites, but are displayed in the various provincial museums storing such valuable and unfortunately portable relics of the past. Some of the finest examples, however, are still on site at a beautiful but unfinished sanctuary of the eleventh century, Prasat Hin Ban Phluang in the north-eastern province of Surin. There on decorative lintels and a pediment, surrounded by stone garlands and foliage of a formal and symmetrical beauty, the crowned and bejewelled god Indra is depicted commandingly mounted on his celestial vehicle, the Erawan elephant (see Plate 3). What is most interesting and noticeable about these Khmer relief-carvings is the fact that unlike the later Thai adaptations of this concept, in every appearance of Erawan, whether with three heads or one, the elephant is depicted as a remarkably realistic, lively, even frisky

[3]While some of these 'palaces', restored from their ruined states by the Fine Arts Department, can today be conveniently visited by tourists (Prasat Muang Singh in the west, and Prasat Hin Phimai, Prasat Khao Phnom Rung, Muang Tham, and Sikhoraphum in the north-east), many outstanding and as yet unrestored examples are accessible only to the armchair traveller through Smitthi Siribhadra's and Elizabeth Moore's splendid book, *Palaces of the Gods: Khmer Art and Architecture in Thailand* (1992).

Fig. 91
Procession with elephant. Lintel at Prasat Narai Jaeng Waeng, Sakon Nakhon Province, eleventh century. (Photograph Pam Taylor)

Fig. 92
Procession with elephants. Lintel above southern corridor entrance to the central sanctuary at Prasat Khao Phnom Rung, Buriram Province, tenth–thirteenth century. (Photograph Rita Ringis)

creature, alleviating the rigid and hieratic symmetry of the scene, suggesting almost a touch of humour (Fig. 93).

No doubt at the time of construction of Khmer sanctuary complexes in Thailand, contributing to the work-force were everyday elephants as draught animals, constant companions of the artisans and builders, helping to haul the massive stone blocks up to the sites and into place. Clearly, these useful everyday elephants provided the models for their celestial cousins carved on the sacred towers. But the elephant, even in a celestial form, is rarely at the centre of action. In fact, as in everyday life, it is generally a helper, an adjunct or associate of the major gods or leading characters whose powers and exploits are alluded to or narrated in the bas-relief carvings of the sanctuaries.

131

Fig. 93
Eastern face of Prasat Hin Ban
Phluang, with lintel of Indra as
directional guardian on an elephant
in profile, Surin Province, eleventh
century. (Photograph Rita Ringis)

Thus the elephant in its celestial form of Erawan is always associated with the god Indra, ruler of Mount Meru and its hosts of
gods. Indra's powers in the heavenly spheres are varied. At Khmer
sanctuaries, he may appear as mere but none the less celestial
guardian of the eastern direction of the universe. However, alluding to his important other role as God of War, he is customarily
and appositely represented riding an elephant, controlling its
prodigious strength. In his additional role as God of Rain he is
the destroyer of the demon of drought, and thus the bearer of
bounty to agriculture. Consequently his celestial mount, the
Erawan elephant, is also linked with the blessings of rain. In fact,
elephants in ancient Indian mythology were associated with rain,
being considered 'the cousins of clouds' (Zimmer, 1960: 160).
Legend has it that long ago, these elephant 'cousins of the clouds'
had wings and could fly. However, a powerful hermit's irritated
curse at their mischievous aviation tricks deprived them of their
powers, rendering them flightless, and since that time, elephants
could be said to be 'clouds sentenced to walk on earth' (Zimmer,
1960: 61).

132

Although this legend does not appear to be widely known in Thailand, the idea of elephants as harbingers of rain appears to have resonated in the Thai consciousness for centuries. In the late seventeenth century, rain-calling ceremonies with elephants as major participants were noted by observers from the French Embassy to Siam at the time (La Loubère, 1693). While various other rain-calling ceremonies are still customary in the north-eastern regions of Thailand, until the early twentieth century, a ceremony similar to that mentioned in the seventeenth century continued to take place. It involved, in addition to appropriate ritual, two elephants in the excited state of *musth*, tethered opposite each other in such a way that they could not harm each other, but allowing for their tusks to meet in this controlled combat, presumably mimicking a clap of thunder. The sound of the tusks engaging gave the ceremony its name, *bamru-nga*, or 'the clashing of tusks' (Giles, 1930a: 67).

Perhaps the association of the many real powers of the elephant with its legendary divine potency contributed to aspects of the elephant being incorporated into depictions of some of the composite 'Animals that never were' that adorn many of the palaces of the gods, and subsequent Thai Buddhist temples. For example, stylized crocodile-like aquatic creatures, *makara*, customarily appear in profile at the entrances to sanctuaries or on antefix decorations and are associated with fertility and abundance. Reinforcing this symbolism, *makara* are often depicted spewing sacred serpents (*naga*), also symbols of plenty. Depending on the time of construction of the particular sanctuary, the *makara* profile might sport either a short snout, or the unfurling trunk of an elephant. The presence of the elephant's trunk in such carvings further emphasizes the idea of fertility and power. Another part-elephant creature adding its talismanic power to a sanctuary was the *gajasimha* (*kotchasingh* in Thai) which combined the body of the lion with the trunk of the elephant (Fig. 94). This doubly powerful creature

Fig. 94
Gajasimha or mythical lion with the trunk of an elephant. Detail of mural from Wat Suthat, nineteenth century. (Photograph Isabel Ringis)

133

is still evident many hundreds of years later in Thai heraldic designs, notably as a component of the standards that decorated the streets of Bangkok during the 1987 celebrations of the sixtieth birthday of His Majesty King Bhumiphol Adulyadej (Fig. 95).

However, on the Khmer towers in Thailand, while these elephant vignettes and sculptural allusions contribute to the overall effect of the magic mountain, they are essentially peripheral to the main sculptures which depict the major Hindu gods Brahma, Vishnu, and Siva. This trinity represents the principles of creation, preservation, and destruction within the universe. Associated with these three major gods are numerous, often contradictory, powers and complementary stories about events in the cycles of cosmic ages of the universe. While in Thailand the gods rarely appear

Fig. 95
Street decorations during the celebrations of His Majesty's sixtieth birthday in 1987. Royal emblem with Garuda at the centre, flanked by *singh* at right and *gajasimha*, lion with the trunk of an elephant, at left. (Photograph courtesy of the Tourism Authority of Thailand)

together as a trinity on the one narrative sculpture, an exceptionally fine lintel at the somewhat unusual early twelfth-century sanctuary of Prasat Hin Sikhoraphum in the province of Surin captures them together and evokes in the devotee's mind some of their celestial duties (Fig. 96). Dominating the scene is the ten-armed Siva Nataraja, who, as Lord of the Cosmic Dance, symbolizes the balance between the forces of destruction and regeneration within the universe. Below the dancing Siva, each seated on a lotus and encircled by garland arches are four graceful figures in concert: his benign and beautiful consort Uma, the four-armed mighty god Vishnu, the four-headed and four-armed god Brahma playing the cymbals, and, playing the drums, Siva's four-armed elephant-headed son, Ganesha. The carving is such that the multiple limbs of the seated gods, not to mention Ganesha's elephant head on a human torso, look perfectly natural.

The origin of Ganesha's peculiar elephantine appearance, as contrasted with the physical grace of his parents, has numerous explanations. One of the most popular concerns a celestial domestic tiff, for while the gods exemplify great powers and exalted virtues, they are often brought low by all too human weaknesses. Long ago, it seems that Siva's beautiful consort Uma/Parvati, irritated by her husband's unwelcome attentions while she was at her bath, created from the scurf of her body a son who was to stand guard outside her chamber. (While Siva, symbolized by a linga, is the embodiment of progenitive powers, in his contrasting role of supreme ascetic or hermit, he practices abstinence; thus his son is not born of actual intercourse.) Such was Siva's wrath at his wife's apparent impudence that his angry vengeful glance at this

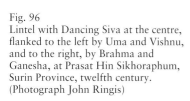

Fig. 96
Lintel with Dancing Siva at the centre, flanked to the left by Uma and Vishnu, and to the right, by Brahma and Ganesha, at Prasat Hin Sikhoraphum, Surin Province, twelfth century. (Photograph John Ringis)

guardian lad decapitated him, much to Uma's distress. To placate his grief-stricken wife, Siva replaced the boy's head with that of a passing elephant, and all was well again. The simplicity of this popular story, however, does not account for Ganesha's enormous and enduring popularity as well as his extensive powers. Providing more satisfying accounts of his origins are various other myths linking Ganesha with the other major Hindu gods. Some of these are alluded to in the carvings embellishing one of the best-known Khmer sites in Thailand, Prasat Khao Phnom Rung.

Set at the top of a commanding hill, construction of this palace of the gods began sometime in the tenth century and was completed in the thirteenth century (see Fig. 89). In recent years, a massive restoration project by the Fine Arts Department rescued it from its romantically ruined but dangerously decayed state, preserving it for present and future generations. To reach its towering sanctuary originally sacred to Siva in his role as supreme ascetic, modern-day Buddhist pilgrims follow a lengthy processional causeway, then ascend the hill by a broad stone stairway linked by terrace platforms. Nearby, to the north-east of the hill stand the remains of a sandstone and laterite pavilion, which the local people have for generations called 'The White Elephant Pavilion', assuming that a king's elephants were stabled here during rituals of long ago, although there is no historical evidence for this. (Smitthi Siribhadra and Moore, 1992: 271).

Reminding the pilgrim that his ascent towards the magic mountain is not only a physical but spiritual journey are stone balustrades carved in the shape of the sacred serpents or *naga*. These are traditionally considered to represent the rainbow, the link between the world of man and the heavens of the gods. The summit attained, the pilgrim passes through the dim encircling galleries and out into the main court of the temple proper, and views the magic mountain in all its carved splendour (see Fig. 90).

Separately depicted on its eastern face are gods in their respective cosmic roles. Barely visible high above on the antefix decoration(s) of the *prang* is a small carving of Indra on the Erawan elephant, in something of a minor role as guardian of the eastern direction of the universe. Well below that, on the large and ornately framed pediment surmounting the projecting eastern main entrance to the central tower, is the sandstone carving of a radiantly handsome ten-armed dancing Siva, eternally poised in momentary equilibrium as Lord of the Cosmic Dance (see Plate 7). Seated below dancing Siva are graceful celestials, sadly defaced by time. Still discernible, however, at the lower right (from the observer's point of view) of the pediment is the god's squat and sturdy son, the crowned elephant-headed Ganesha.

Directly under the pediment of dancing Siva, above the entrance doorway to the sanctuary is perhaps the most famous lintel in Thailand, illustrating primarily a variation of the theme of cosmic destruction and regeneration, but related peripherally to the mythology of elephants and Ganesha. This lintel, restored to the

Thai nation in 1988 after some twenty years of being 'missing' in the West, depicts the god Vishnu reclining on the Ananta-Sesha serpent floating on the Cosmic Ocean, dreaming his cosmic dream, re-creating the universe after its periodic destruction by cosmic conflagration (see Plate 7 and Fig. 35). Emerging on a long stem from his navel is a lotus, seated on which is the god Brahma, the Creator, whose task it is to create in actuality all that Vishnu dreams—everything in the universe. To those familiar with the vocabulary of the myth, elephants are present, though not actually depicted, for one of the myths of the divine origin of elephants is alluded to at this point. It is said that from the many petals and stamens of this miraculous lotus the gods Vishnu, Siva, Brahma, and Agni eventually created the various castes or families of elephants, celestial ancestors of present-day terrestrial elephants.[4]

This same scene of Vishnu dreaming, exalted though it is in terms of creation of a universe, is also peripherally related quite humorously to one of the many myths accounting for the elephantine appearance of Ganesha. In fact, both Vishnu and Siva play the main roles in this popular version of the origins of Ganesha's strange anatomy. Perhaps reflecting this is the fact that images of Ganesha are often depicted holding, among other attributes, either the trident of Siva or the conch shell of Vishnu (Fig. 97). This particular version is closely associated with a custom still occasionally carried out today, the tonsure ceremony. That ritual was customarily practised until the beginning of this century, for the most part in the Thai royal and noble families. According to ancient tradition, until the tonsure ceremony was carried out, the hair at the crown of a child's head was not cut but allowed to grow from the time of birth. It was customarily coiled chignon-like on the top of the head, and encircled by a garland of flowers or even jewels (Fig. 98). The ceremony, of great antiquity (again derived from ancient Indian beliefs and practices), involved the

[4]Nineteenth-century Thai illustrated manuscripts and manuals of design are possibly referring to this occasion of the gods and their creation of celestial elephants in the somewhat baroque depictions and iconography, which depart from the Hindu, of an elephant-headed god, named Vighnesvara, which transliterates from the Sanskrit into the Thai as Phra Phiganet, or the commonly known term in English, Ganesha. These recount a challenge by the god Siva to the God of Fire, Agni, to demonstrate his powers. Fired with enthusiasm, Agni caused flames to burst from both his ears. No ordinary flames these, as at their centre were manifested two celestials, sons of Siva, one of whom was a spectacular being with three elephant heads and six arms. This divinity, called Koncananesvara Sivaputra, translated as Ganesha, son of Siva, held in each of his six hands various white elephant divinities, including the thirty-three-headed sacred Erawan elephant, as well as a less spectacular but no less powerful three-headed elephant called Girimekhala, or 'mountain-cloud'. From these various spectacularly and divinely created elephant divinities, other families of auspicious elephants are said to be descended (Gerini, 1893: 16; Grigson, 1989: 36; Fine Arts Department, 1990). As these baroque variations depart somewhat from traditional iconography, it is possible they were introduced in the nineteenth century to replace traditions of design lost in the destruction of Ayutthaya (Boisselier, 1976: 231).

Fig. 97
Seven-centimetre bronze Khmer-style amulet of the four-armed, one-tusked crowned Ganesha, with a serpent coiled around his upper body. In his right hands he holds a conch shell and his other tusk with which he transcribed the *Mahabharata*, a copy of which he holds in his posterior left hand. Collection of the author. (Photograph Rita Ringis)

ritual cutting of the child's topknot of hair at about the time of puberty, and celebrated the child's safe passage from one stage of life to another, from childhood to maturity. By tradition, at this time, the child thus tonsured was renamed, being known hence-forth with an 'adult' name as against that of its childhood (Gerini, 1893).

However, well before the twentieth century, in the time of the gods, Lord Siva, like any good parent, summoned all the celestials to attend such a tonsure ceremony for one of his sons, a remark-able boy created out of Siva's own chest.[5] All the divinities as-sembled, except for Lord Vishnu, who was taking a nap. Unlike that of mere mortals, Vishnu's nap had cosmic consequences, as it preceded the re-creation of the universe (see Plate 7). Siva's invitation to the ceremony was then reiterated, unfortunately by the distinctly disturbing wail of the conch shell. Understandably, Vishnu, still cosmically dozing, in his half-awakened state, uttered a curse, drowsily complaining about 'a headless brat'. The divine strength of his words, of course, had an undesired effect: the 'brat', Siva's beautiful son, was instantly headless. After much family consternation, Siva commanded the God of Crafts and Arts, Vishvakarman, to scour the universe for a replacement head. But it had to be the head of a being about to die, a being

---

[5]In variations of this myth, the boy is in fact Khandhakumara, Ganesha's brother. However, in popular beliefs, the two have fused (Gerini, 1893: 18).

Fig. 98
A young prince, possibly Prince Chulalongkorn, with a garland of flowers encircling his topknot, indicating that he has not as yet undergone the tonsure ceremony. Engraving from Anna Leonowens, *The English Governess at the Siamese Court*, 1870.

thus facing the west, an inauspicious sleeping direction. Vishvakarman's quest ended when he located an elephant facing west, removed its head, and placed it on the shoulders of the now revived son of Siva, the son who was henceforth to be called Ganesha. Thus, just as Siva's son 'lost his head' but afterwards, as a newly revived being (with an elephant's head, associated with wisdom), was renamed Ganesha, so an earthly child, with his topknot cut off in a ceremony that paid homage to both gods, was renamed and deemed ready to face the future as an adult, presumably with some elements of sense if not wisdom.

This latter story of Siva, Vishnu, and Ganesha is still part and parcel of Thai everyday life. As a result of the inauspicious original tonsure of Siva's son, certain days are considered unlucky for that simplest of activities, a haircut, and even Western-educated Thais may observe that prohibition, perhaps without knowing that it all had to do with a legendary elephant. Furthermore, even today, the direction of the head when sleeping is still considered important. No one will willingly sleep with his or her head to the west, the direction of death, and the direction of the head of the elephant that replaced Siva's son's original human head. And perhaps in memory of the fact that Vishnu, inopportunely awakened from his nap and thus not entirely himself, issued a curse, it is considered very inauspicious to suddenly awaken a sleeping person. Some say that if it is absolutely necessary, it should be done by gently pulling the sleeping person's big toe.

Another version of Ganesha's appearance and origins recounts that after destroying his son's head, Siva sent his celestial mount, the bull Nandin, to scour the universe for a suitable replacement. Unfortunately, the only suitable replacement of the head turned out to be that of the sleeping Erawan, the mighty elephant of the god Indra, who naturally enough refused to countenance the request. Massive battles followed, with Indra aided by the hosts of gods against the might of Siva's bull, which was none the less victorious and cut off the head of the sleeping Erawan. The grieving Indra was commanded to cast the headless body of his beloved elephant into the Cosmic Ocean, in the belief that this paragon of elephants would be revitalized in due course (Doniger O'Flaherty, 1975: 266–9).

Siva then placed this splendid elephant head on the shoulders of his son, who immediately took on a delightful aspect—elephant-headed, short, fat, pot-bellied, and with not two but four arms. Siva's attendants and the defeated gods, rather like the bountiful fairy godmothers in Western fairy-tales, presented various offerings, and named the boy 'ruler of hosts (of gods)' or 'Ganesha'. These offerings, grasped in Ganesha's chubby hands, are his attributes, signifying his powers (see Plates 5 and 6 and Fig. 97). Thus, images of Ganesha may be depicted as holding various attributes, including the elephant goad (bestowed, ironically, by Indra). This is a long handle with a sharp hook that is used by all human elephant riders even today in Thailand to direct an elephant, perhaps to enable it to remove obstacles in its path, as only an elephant can. A noose held by Ganesha links him with the darker destructive powers of Siva but also alludes to the noose that has been used since time immemorial in the capture of wild elephants (see Chapter 7). These attributes, as well as his elephant's head signifying wisdom, make Ganesha particularly popular with those who deal with elephants. Numerous simple shrines are erected in his honour at elephant training camps, as well as in villages in the north-east of Thailand traditionally associated with the care of elephants in captivity.

Images of Ganesha usually have only one tusk, and depict him holding the other, his own, torn out by himself, to use as a writing implement to inscribe the great Indian epic, the *Mahabharata*, as it was being dictated by its creator. For this reason Ganesha is known and venerated as the God of Knowledge and Literature. Thus, aeons since his creation in India, Ganesha as Phra Phiganet in Thailand is the symbol of the Fine Arts Department, and his image is emblazoned at the head of handsome erudite publications as well as on the doors of dusty vehicles that convey staff to the numerous archaeological sites that dot the land.

Another myth about the tusk connects Ganesha with eclipses of the moon, recounting that he tore off his own tusk to hurl at the moon for laughing at him for falling off his mount, the incongruous rat. The latter had reared in alarm at a serpent in its path, causing Ganesha to tumble down and to split open his bulging belly, a result of his weakness for sweets. Not for nothing is Ganesha the remover and overcomer of obstacles, for he promptly coiled the snake around his body, binding up the unfortunate schism, making himself whole again. Thus he is also always depicted as having a serpent draped or tied around his corpulent body.

Seen in their proper context, such symbols which are usually repugnant or unfamiliar in the West—snakes and elephant heads on fat ungainly creatures—take on an endearing charm and meaning once their mythology is appreciated. In fact, paradoxically, contributing to the affection with which Ganesha is universally regarded is his portliness which suggests human weakness and thus renders him more approachable than the sylph-like, slim-waisted gods traditionally depicted in the stage of an eternal golden adolescence.

But what of the poor Erawan who, in an earlier recounted myth, lost his head in the great battle with Siva's bull, Nandin, to provide a head for Ganesha? In the reanimation of the slain Erawan, Vishnu also plays a role. Apart from his cosmic dreaming activities, Vishnu reappears on earth from time to time reincarnated in human or animal form to diminish the powers of evil and disorder. Thus in human form, as Rama in the epic *Ramayana*, he is the virtuous prince who defeats the powers of demons. Another of his reincarnations or *avatara* is in the form of the seemingly lowly tor-

140

toise whose qualities enable him to restore to the world the many precious things lost in the Cosmic Ocean that floods the universe after its periodic destruction by cosmic fire. This exploit of Vishnu is known as the Churning of the Milk Ocean and was occasionally depicted on lintels of Khmer sanctuaries in Thailand (Fig. 99).

In his incarnation as the divine tortoise, Vishnu plunges to the bottom of the Cosmic or Milk Ocean, and there, balances on his back the central mountain of the universe, Mount Meru, to use as a churning stick. On lintel carvings, this is depicted as a central pillar encircled by a serpent, whose head and tail extremities are held by gods and demons in a divine tug of war, churning the ocean in the same way that Indian housewives had for millenniums churned milk (using a common churning stick) to produce butter or ghee, the cooking oil necessary for daily life. In effect, this myth mirrors the mundane activities of daily life, investing them with a divine sanction, for generated in the churning by the gods and demons were many of the everyday necessities of life. Foremost was ghee for cooking, followed by Amrita, the divine nectar or ambrosia bestowing immortality (to be quaffed only by the gods, but sipped, to mankind's cost by the wily demon, Rahu). Also from this milk ocean came Lakshmi, Vishnu's consort as bestower of wealth and abundance, representing the human householder's wife in her careful husbanding of his earthly resources. But of

Fig. 99
The Churning of the Milk Ocean. Detail of a lintel of the eleventh century in Angkor Wat style, at Phimai National Museum. Vishnu in his *avatara* as a tortoise is at the lower centre, with Mount Meru balanced on his back as the churning stick, entwined around which is the Naga serpent, used as a tugging rope by the gods and demons. (Photograph Pam Taylor)

immense importance to the rain-induced fertility of the land, the Erawan elephant arose from the Milk Ocean, unblemished and whole, to become again Indra's mount in the universe.

Thus on Khmer sanctuary towers and later Thai traditional temple buildings, the presence of the Erawan elephant in sculptural form could be read to symbolize the importance of the elephant *per se*, as a beast of burden that contributes to prosperity in the land, and as a supreme example of the 'life-bestowing powers inherent in elephants by virtue of their derivation from the Milk Ocean' (Zimmer, 1960: 161). The symbolic link is clear in the actual name of the elephant. While in common Thai terminology it is 'Erawan', this name comes from the Sanskrit *Airavata* or *Airavana*, which means 'born of the ocean of milk' (*A Dictionary of Buddhism*, 1976: 303).

Just as centuries ago the mystical potency of Khmer sanctuary towers in Thailand was enhanced by the presence of carved stone images of the Hindu gods, their mounts, and their attendants, so today the same gods and their attendants, wrought in coloured mosaic glass or gilded wood, feature prominently and consistently in the sculptural and decorative arts ornamenting the Buddhist temples of Thailand (Fig. 100). Through the continuing prominence of these temple decorations, the Hindu gods have remained part of the folklore and national consciousness. Over the centuries, they have become thoroughly Thai and, with the possible exception of Siva (Phra Issuan), have been closely woven into later Thai Buddhist traditions, being known as Phra Narai (Vishnu), Phra Phrom (Brahma), and Phra In (Indra).

However, the Thai have also further enriched and complicated their traditions by embracing a primarily Buddhist form of architecture, the *stupa*, as it was known in India. From their contacts after the thirteenth century with Sri Lanka, an early stronghold of Theravada Buddhism, the Thai adapted this dome-shaped relic chamber, called *dagoba* in Sri Lanka and *chedi* in Thailand. Unlike the Khmer-derived *prang* which had overtones of temporal power and imperial might, the Sri Lankan-derived *chedi* in Thailand is a purely religious structure, serving to remind the faithful of the Parinirvana of the Buddha, of his death and release from the cycles of rebirth. Furthermore, unlike the lushly decorated Khmer *prang*, the actual bell *chedi* form was rarely ornamented, apart from its basal platforms.

In Sri Lanka, particularly favoured elements of decoration at the bases of the relic domes were large sculptures of elephants. One legend accounting for this practice suggests that when the universe was evolved, from one half of a gigantic egg emerged the Erawan elephant, followed by seven other male elephants. Fortunately, from the other half emerged eight female elephants. These ancestral elephant couples became the vehicles or mounts of guardian deities of the eight directions of the dome-shaped universe, hence their popularity as decorations at the bases of stupas. In fact, in Indian mythology, elephant atlantes were regarded as supporting

142

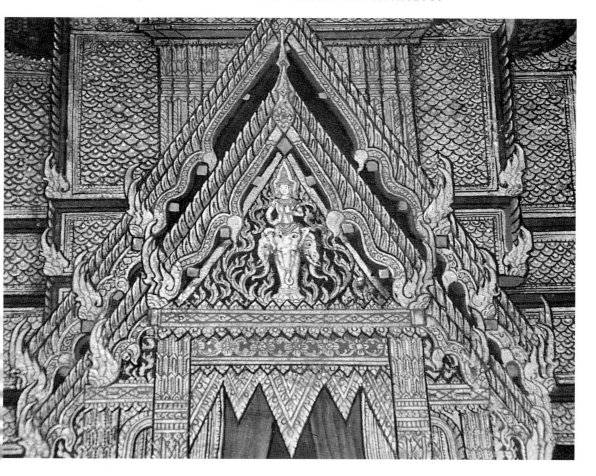

Fig. 100
Detail from a nineteenth-century mural
painting depicting a temple pediment
with Indra on the three-headed Erawan.
Ramakian murals in the galleries
surrounding the Temple of the Emerald
Buddha. (Photograph Isabel Ringis)

the dome-shaped universe on their backs (Zimmer, 1960: 160).

Interestingly, in Burma, which shares various cultural elements with Thailand, earthquakes were traditionally explained by the belief that the elephant atlantes supporting the universe were said to be shifting and changing the position of the dome of the universe pressing on their weary shoulders. However, this belief is apparently not common in Thailand, where earthquakes are relatively rare.

In Thailand, in imitation of the Sri Lankan model, life-sized sculptures of elephants have been popular as the major decorative elements at the bases of *chedi* at various periods in history from about the late thirteenth to about the sixteenth century (Fig. 101). Springing to mind are Wat Chang Lom in Si Satchanalai and Wat Chang Rob in Sukhothai and in Kamphaeng Phet (Fig. 102). In Chiang Mai, there is the ancient *chedi* at Wat Chang Man, enclosed and supported by elephants. Of a later time is Wat Maheyong in Ayutthaya. In most cases, the elephants remain only in fragmented ruined form, or have been severely restored (Fig. 103). Of all these, the earliest, possibly dating from the late thirteenth century, is at Wat Chang Lom ('the temple encircled by elephants') in Si Satchanalai, a vice-regal town during the Sukhothai

Fig. 101
The fragmented remains of thirty-six life-sized brick, laterite, and stucco elephants encircle the lower terrace of the base of the *chedi* at Wat Chang Lom, Si Satchanalai, thirteenth–fourteenth century. (Photograph Paulette de Schaller)

Fig. 102
Projecting from the raised base of the ruins of the *chedi* at Wat Chang Rob, Kamphaeng Phet, are the remains of sixty-eight elephant atlantes. (Photograph courtesy of the Tourism Authority of Thailand)

Fig. 103
Detail of torsos of elephant atlantes with stucco decorations at the chest and legs, Wat Chang Rob, Kamphaeng Phet. (Photograph Rita Ringis)

era. Whatever their significance or date of actual origin, the contemporary effect of these now ruined elephant 'bearers' of the vast terraced and bell-shaped structure can be vividly imagined: thirty-six elephants fashioned in life-size, each separated from the other by pillars that are believed to have supported burning oil-lamps. In the past, no doubt, the effect of the flickering lights of the lamps, as well as the presence of these (fortunately) immobile life-sized elephants, enhanced grand processions by the king and his people in ritual circumambulation of merit-making around the base of this massive terraced *chedi*.

Unlike the earlier Khmer esoteric practices and rituals within the darkness of the stone sanctuaries, confined to the priesthood (Hindu) and their ruler, the Thai Theravada Buddhist practice, as documented from the late thirteenth century onwards, was of communal participation, on a grand scale, in religious ceremonies that were open to all. However, the *chedi* monument was just that, a solid dome structure that cannot be entered, enshrining sacred relics. Thus to provide a communal gathering place for ceremonial, and the ordination of monks, vast rectangular shaped assembly halls were constructed from the late thirteenth century onward. In fact, to Europeans in the present day, these halls are synonymous with the idea of 'temple'. However, to the Thai, the 'temple' is the whole compound of monastic and religious structures and is known by the word *wat*.

Like the earlier highly ornamented sacred sanctuary towers of the Khmer, these Thai assembly halls (*ubosot* and *viharn*) were in time to be lavishly decorated on their outer surfaces. While of

145

these ancient halls only vast pillars remain today, many of the more recent Ayutthaya traditions live on, specifically incorporated in the temples constructed during the foundation of Bangkok, after 1782, and onwards to the present day. These ornate buildings, embellished with gilded carvings, ceramic shards, and glass mosaic, are also in effect palaces fit for the gods, as the Khmer stone sanctuaries had been in the past. Today's equivalent of the Khmer carved stone pediments are the highly decorative triangular gable-boards or pediments set above the main entrances to the Thai assembly halls (Fig. 104). However, unlike the Khmer lintels, the Thai gable-boards rarely display narrative scenes, featuring instead, generally speaking, carved and gilded foliage, and protective originally Hindu divinities. (However, scenes from Life of the Buddha have also become popular in recent times.) For example, the pediments of the ordination hall at the Temple of the Emerald Buddha, in keeping with its royal status, are decorated with the god Vishnu on Garuda. However, Indra on Erawan is present too at

Fig. 104
Gilded pediment with the three-headed Erawan elephant, at the entrance to a monastery assembly hall. (Photograph Rita Ringis)

146

this most important temple in the kingdom on the pediments of a hall visible within the complex but not open to tourists, the Ho Phra Monthian Tham, the Library for Canonical Texts (see also Fig. 69).

Illustrating the strength of these ancient traditions even in the present age is the magnificent ordination hall built during the present reign at the royally sponsored Wat Rajburana at the foot of Memorial Bridge in Bangkok. While the main pediment there is emblazoned with Vishnu on Garuda, the symbol of the present monarch, other pediments feature other gods, including Indra, holding aloft his thunderbolt, seated at the neck of the three very ornately decorated heads of Erawan (see Plate 10). Accompanying Indra are celestials, one of whom holds a noose traditionally used in the capture of wild elephants.

Situated near the familiar landmark of the Giant Swing where ceremonies honouring the god Siva took place until the 1930s is the great monastery of Wat Suthat, from the pediments of which both Vishnu on Garuda and Indra on Erawan preside (see Plate 9 and Fig. 16). This temple was in fact constructed soon after the foundation of Bangkok, at the then magical centre of the city as a propitiatory mandala, to ensure the safety of the new city against continuing Burmese attacks. Thus the great *viharn* represents Mount Meru, and its enclosing galleries replicate the mountain chains of the outer universe. In effect, in the philosophical principles of its layout, it duplicates the Khmer temple prototypes of long ago. Yet, within the halls of this temple, as in all other Thai Buddhist temples, no images of gods are venerated. The decorative and fancifully ornamented celestials remain *outside* the congregation halls; inside only Buddha images preside.

While these images may vary in appearance, depending on the region and date of their creation, in some cases an aspect of their design is unexpectedly related to the subject of elephants. In long-standing traditions of Buddha image-making, certain characteristics (*lakshana*) deriving from poetic concepts of beauty were customarily incorporated in an image to distinguish the appearance of the Buddha. During the Sukhothai period over 500 years ago, the masterly and almost literal expression of these poetic characteristics resulted in images of exceptional spirituality of appearance.

In this stylization of the anatomy of the Buddha, the lids of the gently downcast eyes were said to be like lotus petals; the nose, like the beak of a parrot; the chest, like that of the mighty lion; and the arms were to be as supple 'as the trunk of a young elephant'. These conventions are very clearly evident on Sukhothai Buddha images and because of their obvious appeal to the Thai, elements of that style have continued to be adapted in Buddha images over succeeding centuries well into the present day. To those not familiar with the conventions, the images initially appear distorted anatomically, but these very 'distortions' contribute to the images' serene and other-worldly grace (see Fig. 45).

In any temple assembly hall, on holy days (*wan phra*) devotees prostrate themselves in homage to the teachings, as represented by

Fig. 105
Seated Buddha image in the attitude of
Victory over Mara, at Wat
Suwannaram, Thonburi. Behind the
image the mural depicts the *Traibhum*
or Three Worlds. (Photograph
Kim Retka)

the Buddha image. Seated on the floor in a respectful position, the
congregation listens to sermons and participates in ritual chanting
with the monks. On the walls surrounding the congregation may
be mural paintings which serve to remind the devotees of Buddhist
virtues. For the most part, the murals depict events in the stories
of the Life of the Buddha, and the Former Lives, known as the
Jataka Tales. However, many temples also feature on the west
wall directly behind and partly obscured by the presiding Buddha
image or images, a floor to ceiling rendition of the traditional Thai
Buddhist interpretation of the universe (Fig. 105).[6]

[6]Forming the foundations of that interpretation is the body of illustrated manuscripts called *Traibhumilokasanthan* or 'Maps of the Three Worlds', which in turn
may be based on the mid-fourteenth century 'Sermon on the Three Worlds'
(*Traibhumikatha*, 1987), considered to be the first work of Thai literature. This
latter work describes at length the many ascending levels of the Three Worlds or
planes of existence of mankind, ranging from the lowest hells, through mundane
life to the heavens of the gods, and beyond these worlds of desire and form to
abodes of disembodied bliss, and ultimately to the state of Nirvana. While at the
superficial level this is a poetic vision of inspiring fantasy, describing in detail the
various worlds of many possible levels of rebirth, it is in fact a profound allegory,
providing a metaphorical framework for Buddhist teachings about the spiritual
progress of the individual from the turmoil of worldly desire and delusion to the
state of perfect understanding and thus liberation.

148

Popularly known as the *Traibhum* or 'Three Worlds', this concept is depicted by a series of pillar-like structures set in the billowing waves of the Cosmic Ocean (see Plate 26 and Fig. 105). The central and tallest pillar represents Mount Meru at the centre of the universe. The seven pillars on either side of the central mountain represent in cross-section the seven mountain ranges that encircle Mount Meru at the summit of which is Tavatimsa Heaven. This is a celestial paradise of gardens and pavilions where the Thirty-three Gods pay homage to their ruler, Indra. Also usually depicted here, both in murals and manuscripts, is a pavilion in which resides Indra's white elephant, Erawan, pictured as a majestic elephant with three heads. Illustrated in the manuscripts, however, is an interesting version of the origin of Indra's white elephant (see Plate 15). As animals do not exist at the celestial level of Tavatimsa, whenever necessary, the Thirty-three Gods of the Three Worlds obligingly transform themselves into a white elephant of tremendous size, Erawan, its splendour enhanced by thirty-three heads (*Traibhumikatha*, 1987: 309). Understandably perhaps, except in exceptional circumstances, illustrations of this miraculous elephant generally depict three rather than thirty-three heads (see Figs. 74 and 75).

The manuscripts and mural paintings also illustrate the four continents that lie on the Cosmic Ocean beyond Mount Meru. Our continent or world is called Jambudvipa. There, sheltered in the Himalaya mountains is the mythical Himaphan or Himavanta forest, where many real and fabulous creatures including elephants reside. This is spectacularly illustrated at Wat Suthat, on the pillars in the *viharn*, and in a vast mural in the ordination hall (see Plates 17 and 18). In this fantastic forest, attended by his numerous and playful elephant retinue, a king of elephants bathes in the legendary Lake Chaddanta:

... all the cow elephants flock around him to scrub his body and face and remove dirt and stains until his skin becomes as clear and gleaming white as a conch shell.... After the King of Elephants has finished bathing, the other elephants go down to the lake to bathe themselves. They have fun spraying water left and right. Some thrust their tusks into the ground for play; some amuse themselves by submerging their heads under water ... some gather lotus roots, lotus pods and lotus blooms.... (*Traibhumikatha*, 1987: 407.)

Celestial and mundane elephants of various kinds have also played their parts in the life of the Buddha, as well as in his Former Lives, of which there are said to have been 550. In Thailand, stories of the last ten lives are most popular and are known as the Tosachat or the Ten Birth Stories. These lively parable-like tales, set in legendary Indian kingdoms, illustrate numerous virtues perfected by the Buddha-to-be (Bodhisatta) in his former lives, virtues embodied by noble and exalted characters, but virtues to which even the humblest may aspire, for both high and low life are depicted.

While in mural illustrations of these stories elephants are frequently depicted in a variety of peripheral scenes, in certain episodes they are essential to the tale (and their illustration provides the key to identifying the particular story). Familiar to every Thai, adult or child, is the Vessantara Jataka, exemplifying supreme charity. Its numerous episodes are set into motion when Prince Vessantara is banished with his family to live in the forest, paradoxically for an act of generosity not appreciated by the people of his father's kingdom. The prince is considered to have endangered the well-being of the kingdom by giving away the king's white elephant to a neighbouring but drought-stricken land. Clearly evident here is the perception of prosperity of a land being linked with the possession of a white elephant and its rain-giving powers. In the manuscript and mural renditions of the story, a key scene enabling the observer to read and recognize the Vessantara Jataka is that of the Prince yielding to the entreaties of four Brahmin messengers pleading for the elephant (see Plate 14).

Another tale, the Temiya Jataka, exemplifies the virtue of renunciation of the ephemeral pleasures of a worldly life for a life of mindful contemplation. Illustrations of this depict the young Prince Temiya seated in concentrated stillness, unmoved by the approach of two rogues seated on an elephant which is being incited by well-aimed blows of the goad to trample the prince. Despite its rascally riders, the elephant displays superior understanding and virtue, for it is depicted as resisting their vile encouragement by burying the points of its tusks into the ground, obviously about to tip off the scoundrels as it kneels in homage to the saintly prince.

Always depicted in a key scene from the Mahosadha Jataka are magnificently caparisoned war elephants besieging the walls of a legendary fortified city which is ultimately saved from conquest by the wisdom of a noble adviser to its foolish and gullible king (see Plate 16). Eighteenth- and nineteenth-century mural renditions of such war elephants allow us to visualize the traditional trappings and deployment of the elephant in battles of long ago.

Elephants are also integral in key scenes in manuscript and mural depictions of the Life of the Buddha. Traditionally, a kneeling white elephant holding a lotus in its furled trunk represents the white elephant which appeared in the dream of the slumbering queen, Mahamaya, announcing by its auspicious presence, the birth of a future 'ruler of the world' (see Plate 25). This queen was destined to become the mother of the man who became the Buddha or Enlightened One. It is this episode, the equivalent of the Annunciation to the Virgin Mary in European painting traditions, that should enable a dispassionate observer to comprehend and appreciate the high regard in which white elephants are held in Buddhist countries.

Already described in Chapter 2 is the traditional Thai way of depicting in murals the Enlightenment of the Buddha, when the army of demons led by the Lord of Worldy Desires, the evil Mara on his war elephant, assail the meditating Gautama and are over-come by his superior understanding (see Fig. 21). Occasionally that same episode is rendered in sculpture, an example of which, in bronze miniature form cast in the mid-nineteenth century, is in the collection of the Bangkok National Museum (Fig. 106). However, apposite at this point is a brief 'flashback', as it were, into the past, to fully appreciate the longevity of this tradition in Thailand. The oldest known version of this scene in Thailand is not on a mural, but carved in stone in bas-relief on a lintel dating from the beginning of the twelfth century (Fig. 107). Originally housed in the Khmer-style temple sanctuary of Prasat Hin Phimai, it is now on display in the Phimai National Museum. Still discernible on the weathered stone is the seated Buddha under the bodhi tree, with his right hand placed on his knee in the attitude of Calling the Earth to Witness, thus Subduing Mara, attaining at that moment the Enlightenment of complete understanding of the causes of human suffering and the means whereby to overcome them. On the lintel, directly below the Buddha, depicted facing each other in profile are two mounted war elephants, each attended by followers and dragon-like monsters with elephant trunks, *gajasimha*. This symmetrical alignment of figures represents on the one hand the onslaught of the demons of delusion and ignorance, and on the other, the homage paid by these now-vanquished forces. The lintel is unique, the earliest in both Thailand and Cambodia to represent

Fig. 106
Bronze miniature, 15.5 cm high, depicting Victory over Mara, the scene of Enlightenment, in the National Museum, Bangkok. The bodhi tree and throne below represent the Buddha. To the right is the demon army led by Mara on a war elephant. At the centre, Thoranee, the Earth Goddess, releases the flood from her hair, drowning the demons. To the left, the vanquished Mara on the docile elephant pays homage to the Buddha's virtues. (Photograph courtesy of the National Museum Volunteers, Bangkok)

151

Fig. 107
Stone lintel depicting Buddha's Victory
over Mara, originally from Prasat Hin
Phimai, Nakhon Ratchasima, early
twelfth century, now in the Phimai
National Museum. (Photograph
Pam Taylor)

this scene of the triumph of the Buddha (Smitthi Siribhadra and
Mayurie Veraprasert, 1990: 138).

Legend has it that after his enlightenment, the Buddha preached
his doctrine to ever-increasing numbers of devotees. At one time,
however, troubled by his followers' unruliness, he retired to the
Palileyaka Forest where his simple needs of sustenance were
miraculously fulfilled when a monkey offered him a honeycomb,
and an elephant brought a flask of water (Plate 27). Both in painting and sculpture, this episode of jungle animals kneeling in
homage has an endearing charm. At the same time, the scene is
unusual in that the Buddha is customarily depicted seated on a
rock, as if on a chair in a 'European' manner, with his right hand
extended palm upwards to receive the gifts from these jungle creatures. While this episode is only occasionally depicted in mural
painting and sculpture, the popularity of the story is reflected
unexpectedly as far afield as the highly urbanized city of Singapore
where it appears on the gable or pediment of a Thai temple, the
congregation of which includes many north-eastern up-country
Thai workers labouring on construction sites.

Another favourite scene in Thai Buddhist painting illustrates the
power of concentrated calmness over unreason and brute force.
The Buddha, as any visionary in any age, had enemies who wished
him ill. These sent a wild elephant to cross his path, in the hope of

152

Fig. 108
Ordination procession of villagers with a candidate for ordination on a bedecked elephant at rear. (Photograph courtesy of the Tourism Authority of Thailand)

having him trampled to death. In painting and sculpture, this event is called Taming the Nalagiri Elephant, and is illustrated by the Buddha subduing the elephant's ferocity by the power of his inner strength, represented by his hand raised in a simple gesture of pacification (Plate 28).

At one stage of his life, the Buddha is said to have withdrawn from our world to visit Tavatimsa Heaven, the blessed abode of the god Indra and his white elephant, Erawan. There the Buddha preached to his mother, who had died seven days after his birth, but now, because of her accumulated merit, resided in this celestial paradise. This period of withdrawal in the Buddha's life corresponds in Thailand today to the annual retreat of the monks into scholarly seclusion in the monasteries. This period is also the favoured time for young men to withdraw from their careers and cares of the world, if only once in a lifetime, to be ordained into monastic life for a brief time.

During ceremonies before the actual ordination, every young man re-enacts the life of the Buddha, who was born as a worldly prince and who put aside princely power and a life of ease to become a pilgrim in the quest for liberation from *dukka* or life's inherent unsatisfactoriness. Today, over 2,500 years after the death of the Buddha, young Thai men don white robes with golden decorations (representing the life of the prince) and, after appropriate ceremonial, discard these to put on the saffron robe of

153

poverty, after which, shaven-headed and barefooted, they enter the utter simplicity and austerity of monastic life. While the process sounds rather grim to a European observer, in fact in Thailand, the ordination of a son or brother is considered a most joyous occasion, and festivities abound at this time of year.

In some regions, the candidate for ordination, dressed as a prince, may be carried to the temple on the shoulders of his jubilant friends. In Si Satchanalai district, young men lavishly bedecked in floral garlands, and, in recent times with fashionable dark glasses and make-up, ride richly decorated elephants hired at great expense for the occasion. This tradition of riding an elephant to the temple in effect commemorates in the eyes of the local people the importance of the elephant in the traditions of Buddhism (Fig. 108). In honour of the occasion, the temple grounds may be decorated with handwoven cotton ceremonial cloths in the form of long narrow banners or flags, and the elephant itself may have cloths placed on its head and back as 'saddle' cloths. Woven lovingly by mothers and sisters for the event, these plain handspun and woven cloths and banners are ornamented in a variety of geometric designs, produced by a coloured supplementary weft, in alternating rows of distinctive motifs, including ceremonial offering bowls, horses, temples, trees, mythical animals, birds, and of course, elephants in one form or another (see Fig. 84).

In fact today, as in the past, in any of the variety of media in which Thai craftsmen and women excel, the portrayal of the elephant is not merely decorative but also symbolic of some aspect of Buddhism and its associated mythology, a mythology that abundantly animates the ceremonial as well as everyday life of the Thai people.

7 Hunting, Training, and Working

THE Elephant Round-up held every November in the north-eastern province of Surin is a spectacle to which tourists flock from all over Thailand and the world (see Plates 12 and 13). Carefully choreographed, the Round-up has extensive antecedents, and helps keep alive, if only in enacted spectacle, long-standing traditions, reminders of great elephant hunts of the past (Fig. 109). Purists may decry the commercialization of the Round-up, but then, commerce, not to mention war, and elephants had been traditionally linked throughout the history of Thailand. While the earliest references to hunting and trade in elephants date from the thirteenth century, as recently as the early twentieth century this conjunction was still celebrated in the emblem on the royal Thai government's official seal of the forerunner of the present Revenue Department, always a most important government instrumentality. This seal depicted two kneeling elephants paying obeisance

Fig. 109
Re-enactment at the Surin Elephant Round-up of elephant hunters equipped with hunting lassos and ropes. (Photograph courtesy of the Tourism Authority of Thailand)

Fig. 110
King Udena, legendary tamer of elephants. Facsimile of emblem on Revenue Department tax stamps. (Courtesy of the Tourism Authority of Thailand)

and presenting offerings of forest products to a noble figure holding a lyre. The figure represented King Udena (in Sanskrit, Udayana of Kosambi), considered in ancient and inherited legend as pre-eminent master in the capture and taming of elephants.

Legend has it that long ago in India, a noble hermit, annoyed by the difficulties caused in his simple forest life by the comings and goings of wild elephants, was entrusted by a celestial with secret knowledge of their control. The hermit received a three-stringed lyre, each string of which when plucked and accompanied by chanting an appropriate mantra caused most useful results. Rather like Orpheus soothing the savage beasts, the possessor of the lyre could soothe elephants, or at least persuade them to flee when they were not needed. However, most welcome of all was the power of the third string and appropriate mantra, causing the leader of any herd of elephants to bow and submit before the owner of the lyre. Eventually, the hermit presented the lyre and revealed its wonderful secret to his apparently indigent stepson Udena, in reality the heir to the throne of the kingdom of Kosambi, enabling the young man, in due course, to claim his inheritance, with not a little help from his elephant friends (Giles, 1932: 141–4).[1]

Linking the present pragmatic world with the mythic past, today's Revenue Department tax stamps (familiar to anyone doing business in Thailand), while no longer featuring the elephants, still display the image of Udena and his lyre whose name and magical powers had been invoked for centuries by elephant hunters during ritual preparations for the hunt (Fig. 110). In fact, for some 700 years prior to the twentieth century in Thailand, the hunting, capture, and training of wild elephants for use in commerce, transport, and war was an acknowledged and of necessity esoteric art which was celebrated from time to time by public spectacles.

Documenting these traditions, in the mid-seventeenth century, European traders resident in Ayutthaya recorded that the general population as well as the royal court were present at the culmination of great elephant drives, where tame elephants used as decoys lured their wild and trusting companions into stockades especially built for such occasions. One such observer, Joost Schouten of the Dutch East India Company, noted in 1671 the Siamese esteem for

[1]Another of Udena's legendary adventures is strongly reminiscent of the story of the Trojan Horse. Jealousy of King Udena's love of the hunt and prowess with elephants prompted a rival king to capture him, by fashioning an elephant made of wood, not susceptible to Udena's musical charms, and concealing many warriors within it to ambush him. But the captive Udena, even under threat of execution, could not be persuaded to reveal the secret of his power over elephants. In the course of his imprisonment, however, Udena, out of compassion, revealed the secret to an ugly hunchbacked woman, who was in fact the king's beautiful daughter in disguise. She had been entrusted by her father to betray Udena and to win the secret formula for him. Parental trust was soon forgotten, as Udena and the princess fell in love, and fled in due course to Udena's kingdom on the back of the rival king's famous flying elephant (Giles, 1932).

elephants and their usefulness, noting that the king had more than 3,000 tame elephants in his service, and that 'these creatures . . . are taken in several parts of the country', either in forests by snares or less dangerously but certainly more spectacularly in the stockades or kraals built for the purpose. In the latter method, he noted that some fifteen to twenty tame elephants with their attendants were sent as decoys into 'the wilderness' to lure the wild elephants who 'upon sight of them [the tame elephants], associate with them' and are thus enticed into the kraal enclosures. There, to weaken and disorient the wild elephants, they are 'vexed and tormented with all manner of inventions to make them angry and furious' by wily men dashing at them with lances and goads. 'When the elephant with his running, turning and winding seeks to revenge himself on his tormentors', the men slip to safety through the pillars of the outer or inner enclosures which are spaced widely enough for a man but not an elephant (Caron and Schouten, 1671: 136–8).

One such historic enclosure, looking rather like a tropical Stonehenge, still stands today (Fig. 111). Vast, silent, and deserted, it lies on the fringe of rice fields, some distance from the ancient city of Ayutthaya. Known as the Elephant Kraal, in size it resembles a football field, and is enclosed by raised earthern rampart-like terraces at the lower inner part of which are embedded double rows of tall and massive weathered teak pillars, their austere

Fig. 111
Elephant kraal at Ayutthaya. Engraving from Henri Mouhot, *Travels in the Central Parts of Indo-China* . . . , 1864.

lineality enlivened by finials carved in rotund lotus-bud shapes.

Here, in 1906, one of the last great elephant drives came to its grand conclusion, watched by vast crowds of the population standing on the man-made terraces encircling the stockade. Presiding over the spectacle was King Chulalongkorn (Rama V) and his state guests, including members and representatives of European royalty. The dignitaries and high government officials were accommodated in a temporary pavilion grandstand safely beyond the teak perimeter within which the wild elephants had been lured and driven. Though an occasion dignified by royal protocol, the event would have been somewhat tumultuous to say the least. It is not hard to imagine the billowing dust, the trumpeting and bellowing of the stampeding beasts augmented by the roar of the holiday crowd encouraging the taunting and provoking of the enraged elephants by lithe elephant drivers. These continually slipped beyond their captives' rage through the narrow spaces between the 'elephant-proof' teak pillars.

Yet these spectacular public drives were, for the most part, just that, spectacles celebrating the conclusion to months of hunting. Many of the elephants thus rounded up at Ayutthaya were in fact partially tamed prior to the grand spectacle, for much of the actual hunting, fraught with considerable danger, was done in the forests and jungles, the habitat of the elephant, far away from human settlements. When it is recalled that one fully matured bull elephant has the strength of about seventy men, it would seem foolish if not mad to drive hundreds of wild elephants close to human settlements. In fact, then as now, the great Elephant Drive and Round-up was an overt demonstration and celebration of the skills of the elephant men in 'harvesting' this abundant natural resource.

One such round-up is amusingly described as an 'elephant ball' by a nineteenth-century visitor to Bangkok, the youthful Marquis de Beauvoir:

Nothing can be simpler or more ingenious than the sort of elephant's ball which helps to catch whole herds of these animals in the forests. A certain number of tame elephants are set at liberty; each gentleman gallops off into ... the jungles and invites several wild ladies to dance with him, who follow him with delight; each liberated lady does the same, and brings back a considerable number of dauntless and eager partners. They all return at a wild gallop into an enclosure of which the pailing is made of impenetrable trunks of teak-wood, and into which the clever decoys have led the way; hardy Siamese men then throw lassoes between the legs of the wild elephants, and fasten them by strong ropes to the trees. Taken in the fatal trap of betrayed affections, the unhappy beasts are put on a strict diet, until enfeebled, powerless, and faint from hunger, they submit to a yoke of which food is the reward, and at the end of a year the savage monster of the woods blindly obeys a driver of ten years old. (Beauvoir, 1870: 24.)

While light-heartedly encapsulating some of the major events of a hunt, this description was far from the lengthy, often tedious, and dangerous reality of the expeditions to capture wild elephants. Such expeditions were major operations lasting several months at a time,

158

involving scores of men, tame elephants, and supplies, and many were carried out under the aegis of the Royal Elephant Department.

In the early twentieth century as the modernization of Siam progressed, the practice of these great annual elephant drives fell into gradual decline. To ensure that the traditions underlying these drives did not entirely fade from memory, they were documented, ironically, by one of the very agents of modernization. Francis H. Giles was a distinguished European adviser to the Siamese government, honoured for his role in the foundation of the Revenue Department by the title of Phya Indramontri Srichandrakumara (Fig. 112). Between 1930 and 1938, Giles was also a Vice-President of the Siam Society under Royal Patronage. This learned society was founded in 1904 by Thais and foreigners devoted to

Fig. 112
Francis H. Giles, President of the Siam Society, 1930–8. Also known by the Thai title of Phya Indramontri Srichandrakumara, his rank is indicated by the formal robes and orders which include the Order of the White Elephant, visible on the diagonal sash. (Courtesy of the Siam Society)

159

Fig. 113
Emblem of the Siam Society, devised in
1926. (Photograph Rita Ringis)

promoting an informed and scholarly understanding of Thai history and culture. The Siam Society's emblem and symbol, used on its letterheads and in sculptural relief in the buildings of its premises, comprises the head of an elephant bearing a garland in its trunk (Fig. 113). Encircling the design is a decorative motto in Thai, 'Vicha yang hai kert mitraphap', proclaiming that friendship arises from the pursuit of knowledge.

Between 1929 and 1932, Francis Giles recorded his observations and commentaries on practices of elephant hunting, traditional beliefs about elephants, as well as the many complex and ancient rituals and animistic observances customarily carried out by hunters of elephants. These were revealed to him by master craftsmen of this ancient profession. The substance of much of the information below is a brief interpretation based upon a reading of Giles's lengthy writings in the prestigious *Journal of the Siam Society* (Giles, 1930a, 1930b, and 1932).

According to Giles, traditionally in old Siam, there were two methods employed in the hunting and capture of elephants: one that captured whole herds by driving them into stockades and one that concentrated on separating elephants from their herds, and capturing them individually. The former, known as 'keddah' or 'kraal', derived its name from that of the originally Indian name of the enclosure constructed in the actual forest-hunting areas, far from human habitation. Elephant capture by the kraal method was apparently most common in the northern provinces (Chiang Mai, Chiang Rai, Chiang Saen, and Nan), as well as in the southern seaboard areas, both regions being heavily forested in former times. The second system of elephant hunting, customary in the north-eastern areas, such as Khorat (Nakhon Ratchasima) and Petchabun, involved the tracking of herds of elephants, and their capture by lassoing of individual elephants separated from the herd, with the assistance of trained elephants and their mahouts.

However, while hunting methods differed from region to region, Giles noted that the hunters' sense of participation in a mystical undertaking remained constant. They clearly perceived that they were involved in a venture that was at once extremely physically and spiritually dangerous. The hunt was in effect not only a contest of puny man against the might of the elephant but also an encounter with ancient occult forces and powers to be propitiated within the forests.

As long-standing traditions of animistic beliefs are still strong in Thailand at the end of the twentieth century, how much more so in the past, in the minds of the humble though disciplined men who sought out elephants to capture, for were not these beasts descended from the gods themselves, and was not the forest a place sacred to mysterious and all-powerful spirits? Thus not only were skill and courage in the hunters prerequisite, but 'the discipline of religion and the exercise of virtue' were essential to the undertaking (Giles, 1930: 67). In the hunting of elephants, as in any major undertaking in traditional Thailand, propitiatory rites

160

and ceremonies of ancient origin were observed invoking super-natural help to effect an auspicious and safe outcome (Fig. 114).

In his lengthy observations, Giles records many of the cere-monies in detail. These took the form of prayers and chanted mantras, sacrificial offerings and the divining of omens from these, as well as the use of a special 'forest' or taboo language (depend-ing on the region) while on the hunt. Giles' speculations on the origins of these 'languages' and their psychology are most intri-guing, as is his comment that all words of command used even today in the training of elephants in Thailand (of whatever region) are of foreign origin, clearly indicating the antiquity and inherited nature of elephant lore. Thus it is no surprise that elephant hunters, for the most part, were members of hereditary and hierarch-ical groups, passing their secrets on to those considered worthy.

In the north-eastern Khorat Plateau, for example, would-be hunters started literally from the bottom, as menial assistants to the mahout or rider in charge of an elephant. These menials had to do all the dirty work, until their increasing mastery of various skills, including the secret language of the forest promoted them to the position of 'the mahout of the left' and eventually the 'mahout of the right' within hunting teams. After participating in successful hunts, the latter would be accorded the status of *khu*, and beyond that, *pakam*, the latter entrusted to convey instructions to the

Fig. 114
Re-enactment of a propitiatory ceremony before an elephant hunt, with elders honouring the spirit of the hunting lassos. (Photograph courtesy of the Fine Arts Department, Bangkok)

mahouts from the chief hunter, who was several rungs of responsibility above them, and was known as *patiyai*. The promotion of each man to the various grades within the hierarchy depended, sensibly, on merit, that is, on success (and obviously safety) in the catching of elephants. For example, before promotion to mahout, a menial hunter had to have participated in the capture of fifteen elephants. However, given the traditionally accepted division of elephants into classes, the capture of one noble elephant was equal to five common elephants and so on. All promotions were accompanied by ceremonial and offerings or 'fees' in the form of specific sums of money, chickens, and bottles of liquor. These became the perquisites of the *patiyai* of a particular team, who also served as the 'priest of the hunters' initiating them in elephant and forest lore, and performing at the appropriate times the various propitiatory rites and placatory sacrifices to the spirits, to ensure a favourable outcome of the hunting season. Above the chief hunter of individual teams in a district was the *patiyai-thoat*, a venerable hunter whom all consulted when necessary.

Traditionally, all major activities in Thailand take place according to the appropriate season of the year, which is divided into lunar months. From approximately December to March, during the cool season, in the appropriate regions of the land, men hunted elephants. At this time of year, elephants would be at their physical peak, so that the hunters would choose only the best specimens for their future roles in commerce and war. In fact, a proverb pithily links this seasonal pursuit with another: 'Du chang hai du na nao, du sao hai du na ron', which, loosely translated, means 'the cool season is the time to catch an elephant at its best, while summer is the best for a girl'. Not long after this hunting period, those same hunters, now in the guise of farmers, would be occupied in equally important tasks, of rice planting, and eventually harvesting. However, it should be understood that such teams of skilled men did not merely hunt for themselves. Given the time-consuming nature of a hunt, and its cost, they formed part of the annual drives supervised by the Royal Elephant Department to replenish the supplies of state elephants for transport and war. In fact, in former times, participation in this service would have been part of the corvée due to the Crown.

At the beginning of the hunting season, it was the *patiyai* who summoned the men to leave their homes and villages and to gather together on an auspicious hour and day to prepare, and in effect 'consecrate' themselves to the hunt. In his teachings, this venerable elder expounded that to survive the coming dangers unscathed, the hunters had 'to approach the matter with clean and pure hearts'. Not only was each hunter bound to virtuous and unsullied behaviour before and during the hunt, almost as if he were entering a priesthood, but his wife, during the time of his absence from home, was also to put aside worldly things, and to live in chastity and purity. Such enjoinments on women included ritually 'correct' ways of running their households and performing necessary chores, as well as prohibitions on quar-relling with neighbours or beating the children. Additionally, wives were prohibited from cutting their hair or adorning them-selves with cosmetics and perfumes. Guests or strangers were not to be received in the household during the period of the hunt. Failure to follow the numerous rules would clearly lead to disas-ter for all.

To a modern eye, these domestic prohibitions seem far-fetched, yet there is a clear logic to them. Careful observance of such ritual in simple domestic tasks clearly ensured safety from accidents or misfortunes in the home. The hunter's mind was to be focused entirely on the capture of the elephant, an extremely dangerous occupation. Secure knowledge of such precautions and virtuous behaviour in his home could only have enhanced the hunter's powers of concentration on the forthcoming hunt.

In areas where the stockade system of hunting was favoured, similar propitiatory rites took place. The kraal or stockade enclos-ures, laboriously constructed for the occasion from available materials (bamboo and teak), lay in suitable configurations of the land and formed ingenious elephant-proof traps, similar in struc-ture (but not splendour) to the ceremonial kraal at Ayutthaya mentioned earlier. Of utmost importance was the auspicious siting of the kraal in the forest. To ensure correct selection of a site, ceremonies and offerings to the gods and spirits sought their pro-tection and permission:

Thy slave begs a place to rest,
A place in the forest midst,
Where he may build a palace, a home for elephants,
O Mahadeva [Great God], thy slave asks for a place to rest,
A place to erect a shrine for sacrifice and offerings,
That thou may'st come and enjoy these things. (Giles, 1932: 177.)

Permission or refusal was received through auspicious (or other-wise) dreams of the chief hunter who also invoked the power of his mentor or teacher, prostrating himself before offerings in his honour. Such ceremonial reverence for teachers, or the teacher of a particular art, is still customary today in Thailand at the outset of any undertaking, be it a performance of the classical dance or a Thai boxing match. In the case of the elephant hunting, the chief hunter would chant:

Respectful invitation to the spirit of my Mentor,
To take the place of honour
And to all teachers, please come and help us gain our end.
I make obeisance with pure intent.
Please come, and by thy grace preserve me from danger and all evil.
(Giles, 1932: 172.)

Giles cites numerous such invocations to the spirits of ancestral hunters, and that of the chief hunter's preceptor, his guru (in Thai, *khru* or teacher), welcoming them to partake of food offerings, and hoping that

the magic of this mantra reaching to the elephants
in forest depths,
Causing them infuriated to become,
Roaming here and there, crying in their agony,
They respond to the call of the Holy Mantra of our Preceptor.
They come! (Giles, 1932: 172.)

In the actual construction of the kraal, even in the cutting of the wood and branches for it, ceremonial was lengthy, Giles notes. (Similar ceremonies are still observed in housebuilding in present-day Thailand by the traditionally minded.) Near the kraal, spirit shrines were erected, and placed in them were abundant offerings of various cakes, fruits, cooked meats and fish, rice flowers, and incense to 'invite the sixteen guardian spirits of the earth' and the four gods who guard the directions of the universe to partake of the delicacies. (This was particularly apt since the guardians of the directions of the universe were traditionally thought to be mounted on eight elephant couples.)

In the chanting of mantras of propitiation, many contingencies were covered by invocation of the names and powers of Lord Rama (an avatar of Vishnu), Lord Vishnu himself, Siva, Indra, Varuna, Agni, and even King Udena mentioned above, as well as spirits of the forest. Pragmatically, such ceremonies were a form of contract, for the gods and spirits, after 'having come and eaten of our feast', were urged to 'please help to drive and urge the elephants the forest palace [kraal] to enter' (Giles, 1932: 185).

Even the queen of elephants was invoked to 'induce thy friends and relatives to enter the excellent Palace', to partake of

spirits, rice, our offerings,
Enjoy the juicy creepers, and the green grass
Tempting in their freshness,
Thereafter bathe and gambol, admire the forest palace erected
for thy pleasure. (Giles, 1932: 187.)

Such 'forest palaces' or kraals varied in shape and size, but common to all was the principle of a wide-mouthed funnel-like trench or corridor, some hundreds of metres in length. Constructed of stakes of bamboo and teak, it gradually decreased in width as it led into the kraal or enclosure itself (Fig. 115). This trench was either man-made, or followed suitable topography of the land. In either case, at regular intervals of the trench, high above it in the trees or on raised platforms camouflaged by branches, hunters kept watch, ready to contribute to the impending dramatic stampede. When a herd of suitable size and composition (in keeping with the size of the kraal) had been located, the drive would begin. Ironically, the natural herding patterns of elephants contributed to their capture. An elephant herd is predominantly a nurturing unit composed largely of females with their young, dominated by

164

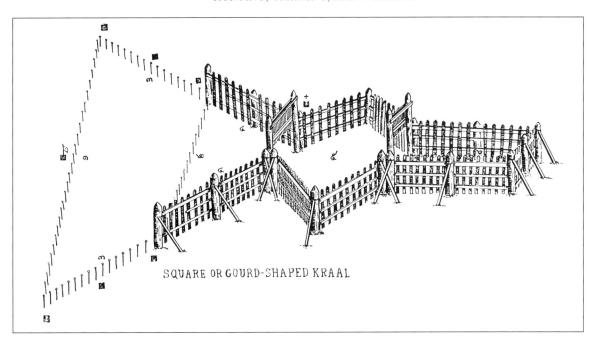

SQUARE OR GOURD-SHAPED KRAAL

Fig. 115
Line drawing of an elephant kraal used in the capture of wild elephants. From Giles, 'Adversaria of Elephant Hunting . . .', 1930a. (Courtesy of the Siam Society)

a matriarch, and only peripherally 'guarded' by bulls, and is thus vulnerable to man's treacherous ingenuity.

During the course of such an engagement, hunters mounted on tame elephants would skilfully manoeuvre the herd overland, sometimes over a period of days until well into this one-way funnel area lined with bamboo and teak stakes. At this point, the cunning silence and stealth of the hunters gave way to explosions of sound: bamboo clackers, firecrackers, and sometimes gunshots exploded suddenly to alarm the herd into stampeding further into this inexorably beckoning funnel. To prevent any retreat, torches and flares were set ablaze on the platforms lining the trench or corridor immediately the herd passed each platform. Waiting patiently at the entrance of the kraal enclosure itself, high above the danger was the gatekeeper, poised to cut the rope that secured the elevated massively heavy drop gate once the herd had rushed in. Crashing down, the stake gate would become deeply embedded into the earth by its own weight, thus confining the animals in the main enclosure.

Here in the kraal, hunters mounted on tame elephants would select the best of the crop (usually the youngest, for purposes of easier training), securing them fast to stakes or trees after some skilful teamwork in lassoing the feet and head (Fig. 116). Those elephants considered unsuitable would be freed at an appropriate time by raising another gate allowing them to flee from the kraal. (Hence the relative 'tameness' of some of the elephants captured time and again each year in various drives, such as the ones at Ayutthaya.)

In the capture of individual elephants and bull tuskers, prized for their usefulness as war elephants, methods differed. For this

Fig. 116
Line drawing of a subdued elephant ready for training. From Giles, 'Adversaria of Elephant Hunting . . .', 1930a. (Courtesy of the Siam Society)

essential and extremely dangerous undertaking, an ingenious decoy method had been devised as described earlier in the seventeenth- and nineteenth-century documentation. While female decoy elephants were used in the initial encounter, of utmost importance in the final capture was the use of lassos. Thus, in the minds of the hunters, these lassos assumed an almost supernatural quality to the extent that these implements came to be considered actual 'spirits', and as such had to be suitably appeased and honoured (Giles, 1930a: 76).

Though the elephant hunters were Buddhist, it was not the Buddha to whom they paid ritual homage. In fact, their rites were a *mélange* of indigenous animistic practices and inherited Brahminical and Buddhist beliefs. Thus the spirits invested in the implements of the hunt (of the lassos, for example), and of the trees and the forest itself, were invoked as well as the various Hindu gods, including Agni, the God of Fire to whom were dedicated the various ritual and utilitarian fires that protected the hunters' camps at night.

In the north-east in traditionally elephant-hunting villages (for example, on the Khorat Plateau), well before setting out for the hunt, sacrifices and offerings were made to the most important implement: the thong or lasso used in the actual capture of elephants. When not in use, these leather ropes, fashioned from treated buffalo hide, were customarily and ritually housed in shrine-like wooden structures, set high off the ground and situated apart from the rest of the village. Great honour was paid to the lasso, for on its strength depended the outcome of the hunt, not to mention lives of the hunters. Additionally, divination of omens, of the bones of sacrificed chickens presented by each would-be hunter at the lasso shrine, forecast the future of the hunt, as well as of those who were to participate in it. Then each man chosen for the hunt, armed with his thong-lasso, rode his hunting elephant into the forest, but not before further rites were carried out to invoke protection and permission from Mother Earth, Mae Thoranee, to enter the forest.

Ritual, and thus discipline, dominated every step of the hunt. Even the choice of the base camp site and its disposition in the forest was according to rule. Three sacred fires were to be kept burning near the camp throughout the hunt, and the men performed daily obeisance here to Agni. The ritual fire served a mundane purpose as well, for wild elephants were thus unlikely to approach the men as they slept. Any impure behaviour on the part of a hunter seen as likely to jeopardize the hunt was punished, after confession, by whipping the offender, who was obliged to run around the camp thrice on all fours, trumpeting like an elephant, carrying his lassos on his back (Giles, 1930a: 88).

When all numerous rites and ceremonies had been concluded, the hunt began. Actually tracking a herd might take days, but eventually, through his skill, the chief hunter would lead the men, all mounted on trained elephants, towards a grazing herd, ordering the mahouts of the right and left to position themselves accordingly near the herd, in effect surrounding it. Thus enclosed, the wild elephants would be stealthily and gradually driven towards the chief and other hunters, all of whom were apparently 'invisible' to the herd as they had annointed themselves with the fragrance of wild elephant dung. Inexplicable at this point is the fact that even very close up, wild elephants do not seem to 'see' the men on the backs of their tame companions. Further contributing to the mystery of this 'blindness' is, on the other hand, the often observed fact that trained draught elephants carrying riders or howdahs on their backs, do actually see and apparently estimate their own and combined riders' height to avoid low-lying tree branches, and even at times remove overhead potentially obstructing branches with their trunks.

Many such hunts culminated on moonlit nights. Caught by surprise, the wild elephant herd would be driven into complete disarray and panic by shouts of the hunters on their trained elephants. Taking advantage of the general mêlée, small teams of hunters would choose a particular elephant to capture, and then pursue it closely on their trained elephants. Around the neck of each tame pursuing elephant was secured one end of the lengthy thong-lasso. The other end of this supple thong, the actual circularly shaped lasso, was attached to a pole, held seemingly casually by the mahout over his shoulder, rather surprisingly like a delicate butterfly net might be held. During the jostling pursuit of the chosen one, several riders would lower their poles with their attached lassos to nearly ground level, ready to slip over, tighten, and snare the hind feet of the panic-stricken and disoriented wild elephant (see Fig. 116). If successfully snared, the poles would be dropped, and the lassos would tighten of their own accord around the hapless elephant's leg or legs as it tried to flee its pursuers. At this point, with great force and manoeuvring skill, the hunters seated on the tame elephants would take up the slack of these noosed thongs to immobilize the captive elephant and secure it eventually to a suitable clump of trees. Then would ensue a lengthy and dangerous struggle to place a leather neck-rope over the head and neck of the captured animal, so that it might be completely secured.

Traditional taming of captured wild elephants followed certain procedures that sit badly with today's animal lovers. The selected animal was pegged at all fours and tied to training posts, or lured by proffered delicacies into 'cradles' or 'crushes', wooden enclosures of stout posts in which the elephant is totally confined, barely able to move. Such confinement might vary from a few days to even weeks, depending on the elephant's response to its captors' directions. Such taming depended largely on confusing the animal and in effect breaking its spirit. Confined in a crush, or tied to a tree, it would be deprived of sleep, alternately starved or fed, in turn physically tormented and then rewarded with kindness until it became clear to it that both the removal of pain and the pleasures ensuing came at the hands of its tormentors. Also at all times, one or other of the trained elephants that had assisted (perfidiously perhaps) in the capture of the hapless wild elephant remained close by with its attendant, to help 'teach' the captive by example.

As time passed, the captive was frequently rewarded with delicacies, and it gradually came to accept without struggle the placing around its neck of various ropes, chains, or harnesses, and finally, even the weight of a man on its back. This type of initial subduing was traditional, whether the subject was a wild fully grown tusker, or a baby born in captivity. The spirit of the former had to be subdued and made obedient to man's commands, the latter was to be accustomed to the fact that its days of unfettered play and freedom at its mother's side were over, and that gradually disciplined behaviour was to be instilled.

According to Giles, prior to the hunters' return to the 'civilized' world of man, the chief hunter also celebrated the completion of the contract with the spirits of the forest, declaring with proper ceremony that the captured elephants were now the property of the hunters, and that all the previous offerings to the spirits had certainly compensated them for the capture of these elephants. The hunters themselves also performed rites that freed them from any obligations to the spirits. All contracts thus fulfilled, the hunters removed and burnt the sacred protective cords that had encircled their waists throughout the hunt, and they were now free to return to the use of their daily language and customary behaviour.

Once the captured elephant had been subdued, and in effect resigned to its fate, it would be either placed in government service, or sold at an elephant market. Of course, during this colourful affair, extensive and painstaking assessments of each elephant's age and physical characteristics were taken into account, as these affected the value and potential usefulness of the elephant in whatever field it was to be further trained. Naturally, in former times, noble and handsome tuskers would be reserved for war training, carried out under the auspices of Krom Kochabal, or the Royal Elephant Department.

The more mundane working elephants, destined as pack animals for baggage and transport, were, however, no less valuable. Owners of these pack animals were guaranteed substantial incomes, for apart from riverine traffic, given the then heavily forested terrain, elephants were the major means of transportation overland. In fact, distances between one place and the next were judged by how many days it would take to cover them on elephant back. As recently as the 1920s, Reginald Campbell, then a young teak wallah, recorded in his memoirs of the same name a sixteen-day elephant-back journey from Lampang to his concession forest in Amphoe Ngao, today a mere hour or so by car (Campbell, 1935).

For such journeys, the cost of hire of elephants always included their drivers, and sometimes attendants, as well as the distance and weight of goods to be transported. Holt S. Hallett, author of *A Thousand Miles on an Elephant in the Shan States*, was something of an expert on this by the end of his 1876 journey to reconnoitre for suitable routes for the construction of a railway to transport British goods from Burma through northern Thailand, and possibly on to China. Such a railway would have had considerable impact, for Hallett noted that at the time, the cost of transporation per ton of rice by elephant was fifty-four times that by rail (Hallett, 1890).

Transport elephants were assessed and valued not only for their strength as baggage carriers, but also for the quality of their gait, a very important factor for a long-distance journey. While a handsome long-legged male elephant might enhance the dignified appearance of the rider, more often than not the rider was severely incommoded by its jolting gait (one anonymous but vividly descriptive traveller described the experience as being akin to riding a three-legged cow in a sand-box). Thus, in the hire of transport elephants, rather than mere appearance, their ease and speed of gait fetched higher prices. Ruefully, Hallett confessed that female elephants were actually the more comfortable for long journeys, but riding females was considered ignoble for a man of substance. Indeed, Carl Bock, Hallett's predecessor in recording his lengthy journeys in Siam, apparently committed that solecism. Tricked in the far north by a Laotian chieftain in a 'bad elephant deal', he engaged a female elephant to ride, and became the subject of jibes and insults even from up-country yokels. 'A she-elephant, indeed, holds much the same place in the estimation of the Siamese and Laosians as a donkey does in England, and a

169

gentleman riding down Rotten Row mounted on the humble ass would hardly be exposed to greater ridicule.' (Bock, 1884: 252.)

However, no such ridicule was directed against the female elephant in the ancient role of seductress and skilful decoy in the art of hunting. In war her feminine presence was necessary at the side of male war elephants, as observed in the seventeenth century by La Loubère, who noted that the Siamese considered the presence of two female elephants beside each male war elephant as not only necessary for the latter's 'dignity' but also that 'it would be very difficult always to govern the males without the Company of the females' (La Loubère, 1693: 92).

Nor was the female's contribution ignored in the development of a new 'art'—that of working as part of a team in the extraction of the 'green gold' of the forests. With the introduction of the timber industry at the end of the last century, and its rapid expansion throughout this century, elephants of all kinds became even more closely associated with commerce and the creation of wealth. Indeed, it is said by some that without the elephant's responsive intelligence and unique ability to work in such difficult terrain, there could have been no teak industry. Throughout this century timber concession companies used thousands of elephants in all aspects of their logging operations. While the companies had numbers of their own trained elephants, by far the largest numbers were chartered *en masse* from village headmen and trader contractors whose 'supplies' originated from the traditional hunting expeditions, but by this time under licence from the government.

Until very recently, the Young Elephant Training School established by the Forest Industry Organization at Lampang conducted long-term training programmes to prepare elephants for the logging industry (Fig. 117). There, young calves would romp free while remaining closely attached to their mothers until they were fully weaned at about the age of three years. At that time, the baby was lured by its love of delicacies into the pen or 'crush' and confined there, much against its will, for several days to prepare it for obedience training. During this time, it would become acquainted with its mahout, and by a process, of alternate teasing, cajoling, smacking, and rewards of delicacies for appropriate responses, it would learn to bow to the will of the mahout, who was to be its master, generally for life.

While the calf is not actively ill-treated, it is evident that such unexpected confinement is distressing and painful. Indeed, a fourteenth-century text compares the pain of childbirth to this process, commenting that 'the baby feels great pain in its body and its flesh at this time, as if it were an elephant being pulled and pushed through a too narrow gate, so narrow that its body can pass through only with great difficulty' (*Traibhumikatha*, 1987: 133).

To alleviate the calf's obvious distress at the separation from its mother, traditional animistic ceremonies were carried out by mahouts under the leadership of a mahout of exceptional spiritual qualities. These rites involved symbolically severing the dependent

170

Fig. 117
Trained elephants in a forest camp.
(Photograph courtesy of the Forest
Industry Organization)

mother–child relationship, after which the mother would return to full-time work without hindrance. The subdued calf, no longer exhibiting distress, would willingly undertake training on a regular basis, according to mahouts interviewed during the course of this research.

Traditionally, according to their mahouts, the training of a young elephant is methodical, repetitive, and not unduly onerous for the youngster, beginning with a daily bath in the nearby river. In time, the calf learns to respond to the simple words of command that will form its working vocabulary for the rest of its life. Following appropriate signals, it learns to walk in single file, in pairs, or in procession, depending on the command. It is also introduced to the training harness and the leg hobbles, and becomes comfortable with their weight, as well as the weight of the mahout on its back. By that time, it will have mastered the movements necessary to assist the mahout to mount or dismount (Amnuay Corvanich, 1976). By the age of six, the elephant would have graduated to 'high school', and until the age of ten, it would attend classes daily for some nine months of the year. Because of the Thai elephant's susceptibility to heat, training is held during the cool hours of the morning only, from about 6 a.m. until 11 a.m., with the afternoon being free for rest and roaming in the forest for fodder. Understandably, vacation time is during the three months of the hot season, from about March to May.

During its secondary schooling, the elephant's time would be spent refining its earlier accomplishments in the more complex tasks of piling logs, dragging logs, carrying or pushing logs up and down hills and into streams, using the trunk and tusks. In addition, its classwork involved becoming accustomed to 'the commotion and noises of various machineries employed in forest working, like trucks, tractors, power chain saws' (Amnuay Corvanich, 1976), and even being trucked from place to place, camp to camp, when necessary. When deemed competent, it would graduate at about the age of ten to tackle the lighter tasks in the forest, and begin the work of its maturity at about the age of sixteen, as part of a team or herd not necessarily related by blood ties. In fact, the working elephant's closest relationship is with his mahout.

Life in the teak camps up-country has been extensively documented by numerous former teak wallahs and the reader is directed to their memoirs listed in the reference section. Common to all their observations is that of utmost importance in teak working was the health and well-being of the elephants in the camps. In fact, the health of the trained elephant in captivity depended on its being kept under conditions as close to the wild and free state as possible. Thus, despite its legendary strength, the trained elephant cannot work all day. Whenever possible, the elephant would work only in the early morning hours, for a maximum of six hours, until about noon, for about four days at a time, and then two days would be scheduled as rest periods. Furthermore, during three to four months of the hot season, when streams ran dry in the concession forests, the elephants would be led up into the higher forest valleys in the mountains for recreation (Marshall, 1959). On their return, they would be assigned to a new area of the forest to be worked.

Reminiscing about his days in the 1950s as a young teak wallah up-country for the Borneo Company, Dacre F. A. Raikes, long-term resident of Thailand and also a recent Vice-President of the Siam Society, recalled that during a period of many weeks within a given forest concession area, he would travel a circuit between some six camps which were scattered throughout the forest, spending three to four days in each. Accompanying him was his entourage which was responsible for pitching his tent and maintaining some of the necessities of civilized life, such as supervising afternoon tea, during which time the health of the elephants could be noted in a suitably relaxed manner after a hard day for all in the forest (Dacre F. A. Raikes, pers. com.).

While the concession company as a whole might own many hundreds of elephants, hundreds of others were chartered from village headmen, who were responsible for the health of their own elephants. Company elephants, however, were the responsibility of the teak wallah, between five and ten in each of his camps, and these were inspected daily, with the details of their physical condition recorded in each elephant's individual logbook. Daily inspection was essential, so that any potential problem could be immediately arrested, for once an elephant was sick, it could be off the work team for weeks, resulting in an enormous loss in productivity, not to mention distress for the elephant.

Dacre Raikes recalled the elephants 'pirouetting' before him at teatime, refreshed from their well-deserved bathing and playing in the forest streams. Most important to note was whether the elephants appeared in overall good and frisky health. Their eyes were checked, the proper flapping of their ears noted, and the skin on their backs and bodies was scrutinized minutely for any spots worn down by a poorly fitted harness, for these areas could develop into sores or abcesses. The feet and toes were also examined for any slight injury sustained during the day's work. Any injuries or abcesses required laborious care, and often the painful remedies became the reluctant duty of the teak wallah, who as a result was generally regarded by the elephants with a wary eye. If all was well, duly noted in the logbook would be the 'three Fs'—fit, fat, and free from sores. The chartered elephants were just as carefully monitored by their private owners, and any lack in the mahout's care for his elephant was severely punished. For the most part, the elephants' keepers were very solicitous of their charges, as it was in everybody's interest to keep the animals in good condition.

The number of elephants in the camps at any given time depended on the size of the forest area being harvested. This was estimated by the teak wallah after several forays from his base camp in the forest. He would select trees suitable for culling, estimate their size and tonnage, as well as distance from the river, railhead, or roadhead of the particular concession area to be harvested. Lengthy negotiations would then begin with the elephant contractors who would in their own good time accept so much per cubic metre to fell, saw, and drag the timber to the river or other point of transportation. Always included in the cost of hire for the whole season was both mahout and elephant, for the close partnership between a particular man and his beast was at the core of success in such laborious and often dangerous work.

Indeed, the accepted dangers of the hauling of heavy logs in rugged terrain were sometimes compounded by deliberately continuing to employ an elephant in a state of *musth*, or by keeping on the odd naturally bad-tempered elephant, recognizable by his habitually 'stroppy' aggressive gait. This was done by owners unwilling to lose productivity time, and thus part of their fee. In those cases, according to Dacre Raikes, the mahout was paid extra, and was generally willing to carry on as a matter of pride in his capacity to control his elephant.

In the work of lifting and pushing of logs, the tusked male elephants were of invaluable use (see Figs. 38–40). However, in the case of known and continually aggressive male elephants, the tusks were usually tipped or cut off to prevent their harming not only the men but other elephants working with them. This was done in a way that was not painful, and in fact, the tusks would continue to grow again. Care was taken not to tip too close to the nerve, during which process the tusk being tipped was continually sprayed with water to eliminate the build-up of heat and therefore potential pain.

Tusked or tuskless, the elephants directed by their mahouts and supervised by the teak wallah laboured not only in the rugged terrain of the forests, but also on steep river banks hauling the logs into the river, directing masses of logs into the current. The elephant's great swimming skills were put to particularly good use here, as teams of elephants, with extraordinary skill, intelligent persistence, and strength, would disperse massive log-jams, freeing the flow of teak to proceed to the major collection points downriver. While this riverine transport was not only traditional but also cost-effective, the rainy season could play havoc with the plans of men and elephants. Dacre Raikes recalled many such occasions when heavy rains further up-country could turn the waterways lower down into funnels of roaring water, carrying masses of huge logs as lightly as matchsticks, with logs shooting out of the thundering rapids 'like corks popping out of champagne bottles'. That elephants could under these conditions be persuaded to enter the water is testimony to their trust, perhaps misplaced, in man.

From the mid-1950s onward, to ensure tighter control of the transportation of logs, more roads were constructed into the forests. Transport of the teak logs by trucks became common, and, concomitant with increased costs, the price of teak increased, and thus in harvesting the green gold, the process of deforestation accelerated. Since that time, despite advances in modern technology, that ancient beast of burden, the elephant, has continued to provide the best means of efficiently extracting timber from the forests. Though machines are now extensively used for the actual cutting of trees, the elephant's skill in manoeuvring itself with the logs obviates the need to build even more roads into the forest. Theoretically, this protects the forest's integrity, while still allowing access to individual logs marked for felling. While roads into forests meant additional expense to the loggers in the early days of the timber industry, nowadays those same roads into the forests mean greater access for illegal logging. Sadly, in the present-day rush to get rich quick, illegal logging still apparently continues on a surreptitious scale in remote areas. The elephant's skill is again necessary to the predators, for machinery can be heard over long distances, and traced by the government forest rangers, whereas the elephant's steady pace and silent strength can be effectively harnessed by the unscrupulous.

In the last decade of the twentieth century, as Thailand is undergoing a period of unparalleled economic development, traditional ways of life on the land appear to be threatened by complex forces, including considerable and much-debated environmental change accompanying the processes of development. For example, deforested areas and even wetlands have been reafforested by a

Fig. 118
Elephants on Sukhumvit Road,
Bangkok. (Photograph Kim Retka)

rapidly burgeoning industry, that of eucalyptus plantations. While these exotic non-native plants have a rapid and high economic yield, their introduction is meeting with considerable resistance from many quarters, not the least being villagers who complain that in the eucalyptus-growing areas, not even the buffaloes but least of all the elephants can sustain themselves.

Encapsulating such current and seemingly irresolvable problems besetting the authorities is the not uncommon sight of the peripatetic elephant and his keeper ambling in and out of Bangkok's traffic-choked streets (Fig. 118). The elephant is likely to have come from the Ta Klang area in Surin province, born and raised in captivity from stock that was captured generations ago by the Guay people of Surin in their hunting expeditions traditionally carried out in the forest areas along the Cambodian border. Not long ago, the Ta Klang village was famous for its numerous elephants, which formed part of the extensive contingents of forest labourers during the logging seasons up in the north. Today, in the understandable pursuit of a better livelihood and future, these elephants and their keepers, like so many others of their countrymen, are leaving the rural areas, lured to the bright lights of Bangkok, the centre of wealth, understandably but inappropriately in pursuit of a better livelihood and future.

However, given the Thai people's long-standing traditions of innovation and flexibility, it seems reasonable to hope that this conflict between presently declining rural traditions and dynamic development generated from the cities will be resolved to the benefit of both in the not-too-distant future.

175

Glossary

Parentheses indicate, where necessary, linguistic variations of words commonly used in the description of Thai art. (S = Sanskrit; P = Pali; T = Thai.)

Antefix: In the context of Thai art history, a term defining separate and various ornamental decorative elements projecting upward from the multiple horizontal levels of Khmer and Khmer-derived Thai towers or *prang* structures. In the case of the former, these antefixes are highly decorated with carvings; in the case of the latter, they may be relatively plain.

Airavana/Airavata (S): The elephant mount or vehicle of the Hindu god Indra. Known as Erawan in Thai, it is usually portrayed with three heads.

Avatar (T): The reincarnation of a deity on earth, usually in reference to one of the ten incarnations of the Hindu god Vishnu.

Ayutthaya (T), also *Ayudhya*: The capital of Siam or Thailand from 1350 to 1767. The name derives from the Sanskrit *Ayodhya*, the city of Lord Rama in the Indian epic *Ramayana*.

Bodhisatta (P): In Theravada Buddhism, this term is used to describe the Prince Siddhartha Gautama before his Enlightenment as the Buddha, and also refers to the central character in each of the Jataka Tales, stories of the previous lives of the Buddha.

Bot (T): *See Ubosot.*

Brahma (S), *Phra Phrom* (T): In the Hindu Trinity, Brahma is the Creator, along with Vishnu the Preserver, and Siva the Destroyer. In Thai Buddhist sculpture or painting, Brahma is depicted as four-faced and four-armed. In the depiction of Thai Buddhist cosmology in painting, the Upper Heavens are inhabited by multitudes of divine spiritually evolved four-faced beings, also called Brahmas.

Brahman (S), *Brahmin*: A member of the priestly class in India. In Thailand, white-clad, black-haired Brahmin priests still officiate at various court ceremonies that have their roots in pre-Buddhist Indian traditions.

Brahmanism: The ancient religion of pre-Buddhist India, from which Hinduism evolved.

Buddha: The Enlightened One who has attained perfect understanding of the causes of human suffering, and thus the means of release from future rebirths.

Chang pheuak: The 'white' elephant, whose special characteristics are numerously exclusive, making the elephant exceptional. By law, such elephants must be presented to the Crown.

Chang samkhan: The general term for auspiciously significant elephants. These may include any elephants of especially unusual aspect. They are usually presented to the Crown.

176

Chedi (T): Buddhist relic chamber in Thailand, usually in a domed or bell-shaped structure.

Deva (S, P), *Devata* (T): The celestial inhabitants of one of the lower heavens of Buddhist cosmology; frequently depicted in Thai murals in worshipful posture as observers of various events in the life of the Buddha.

Erawan: The Thai name of Indra's three-headed elephant mount or vehicle.

Gable-board: In Thai religious architecture, the recessed triangular face of a building between the two slanting roof eaves delineated by the barge-board; usually ornately carved and depicting Hindu gods in Thai form. *See also Pediment.*

Gajasastra (S), *Kotchalakshana* (T): Manuals of elephant lore.

Gajasimha (S), *Kotchasingh* (T): Mythical creature with the body of a lion and the trunk of an elephant.

Ganesha: The elephant-headed god, and son of Siva. Honoured as Remover of Obstacles and, in Thailand, as the God of Wisdom and the Arts.

Garuda (S), *Khrut* (T): The king of birds, a mythical half-man, half-bird vehicle of the Hindu god Vishnu. While Garuda is the vehicle of the god, many lesser *garuda* inhabit the mythical world.

Gautama (S), *Gotama* (P): Part of the given name of the historical Buddha, who was called 'Siddhartha' as a prince, 'Gautama' when he became an ascetic, and 'Buddha' when he achieved Enlightenment.

Himaphan (P, T): A mythical forest inhabited by real and mythical creatures, depicted in Thai painting and decorative sculpture. Elephants of many colours live there.

Hinayana (S): Though nowadays frequently applied interchangeably with the term 'Theravada' to the form of Buddhism practised in Sri Lanka, Thailand, and parts of mainland South-East Asia, this is a somewhat derogatory term meaning the 'Lesser Way' or 'Lesser Means of Progression' towards liberation from the cycles of rebirths, as against Mahayana (the 'Greater Way') espoused by the Mahayana Buddhist sects. *See also Theravada.*

Hinduism: The predominant religion of modern India, derived from ancient Brahmanism.

Indra (S, P), *Phra In* (T): A powerful god originating in ancient India, but popular (as *Sakka*) in Thailand as a guardian of Buddhism. Indra was the God of War, and the God of Rain. In Thai art he may be represented wielding his principal weapon, the *vajra* or thunderbolt, and his bow, representing the rainbow. Indra rides the celestial white elephant, Erawan, and is the ruler of Tavatimsa Heaven on the summit of Mount Meru. He is also the guardian of the eastern direction of the universe, and may be featured in sculptural form on the eastern faces of religious structures.

Jataka (S, P): A 'birth story' referring to one of the 550 previous lives of the Buddha, collectively known as the Jataka Tales, in each of which a particular virtue is practised to perfection. Popularly depicted in Thai mural paintings are the last ten lives, known as the Tosachat, or Ten Birth Stories.

Keddah: An enclosure or stockade used in the hunting and capture of elephants. *See also Kraal.*

Khmer: Used either as a noun or adjective, referring to the people of Cambodia, both ancient and modern. Based in Angkor, the Khmer empire held sway over much of Indo-China from the seventh to the fourteenth centuries AD.

177

Kraal: The stockade in which elephants traditionally were captured. *See also Keddah*.

Linga (P, S): The phallic emblem and symbol of the Hindu god Siva. Cult object venerated in Khmer Hindu temples.

Lintel: In Khmer temples, the rectangular stone cross-beam above the window or doorway, directly below the pediment. Usually carved with narrative scenes.

Mahout: The keeper or rider of the elephant.

Makara (S, P, T): A mythical aquatic crocodile-like composite creature, deriving from India, variously depicted in religious architectural decoration in South-East Asia, usually at the lower edges of arch-like entrances, as a symbol of abundance.

Mara (S, P), *Marn* (T): The demon personification of worldly desires and delusions, assailant of Gautama during his meditation prior to becoming the Buddha.

Meru, Mount (S, P), also *(Su)meru, Phra (Su)men* (T): In Hindu and Buddhist cosmology, the axis or mountain at the centre of the Universe; abode of the fabulous creatures of Thai decorative arts and painting. Indra's Tavatimsa Heaven is at the summit. In Thailand, a focal structure within a temple, a temporary structure for a cremation, or even an artificial mountain created for ancient royal ceremonies, may be considered to be a replica of Mount Meru.

Mudmee (T): The Thai term (*ikat* in Malay/Indonesian) for a technique of distinctive textile decoration. A wide variety of patterns, most of them traditonal, is produced by variously tying and dyeing groups of yarn at specific predetermined intervals before their being woven into cloth, either cotton or silk. This style is much favoured in the north-east of Thailand.

Musth (T): A periodic state of destructive and violent behaviour in mature male elephants.

Naga (S, P), *Naak* (T): Semi-divine serpent-like creatures which can assume human form at will; symbols of abundance and fertility as guardians of the earth's waters, be they river, pond, or cloud.

Nirvana (S), *Nibbana* (P), *(Maha)parinirvana* (S): The former two variations of the term refer to the state of perfect equanimity and understanding, and thus release from suffering, delusion, and future rebirths, an indescribable condition attained upon Enlightenment, while still being earthbound. The latter term refers to the perfect release attained at the time of Death of the Buddha.

Pali (S, P): An ancient Indo-Aryan language used by the Theravada sect of Buddhism in writing its scriptures. Pali is still the sacred language used today in Theravada Buddhism in Thailand.

Pediment: On a Khmer monument, single or multiple pediments surmount the rectangular lintel which is above a doorway or window. A pediment may consist of two parts: a decorative, sometimes sinuous, arch shape that in turn delineates and encloses a somewhat triangular area usually containing a carved descriptive or narrative scene. *See also Gable-board*.

Phra (T): Honorific, meaning 'Venerable' or 'Lord'.

Prang (T): A tower-like form of Thai Buddhist relic chamber, derived from the Khmer. The focal structure at Wat Arun, the Temple of the Dawn, is an example of a *prang*.

Prasat (T): The term, meaning 'palace', is applied to both royal and religious architecture that features on its roof-line a tiered spire-like tower or pyramidal finial(s).

Rama (S), *Phra Ram* (T): The hero of the Indian epic *Ramayana*, and seventh incarnation of the god Vishnu.

Ramakian, Ramakien (T): The Thai version of the Indian epic *Ramayana*.

Ramayana: The ancient Indian epic of the adventures of Rama.

Ravana (S), *Tosakan* (T): The ten-headed demon leader, enemy of Rama in the *Ramayana*.

Siva (S), *Phra Issuan* (T): A major god of Hinduism having many forms, some terrifying. In the Hindu Trinity, Siva is the Destroyer. Paradoxically, Siva is also venerated as the God of Fertility, in the form of the *linga*. In Khmer and Thai art, as against Indian art, the god is generally depicted in benign and mild forms.

Siddhartha Gautama (S), *Siddhatta Gotama* (P): The given name for the prince who became the historical Buddha.

Singha (S), *Singh* (T): Lion, usually depicted in its mythical form.

Stucco: A type of plaster used in architectural ornamentation and sculpture.

Stupa (S): Originally a burial mound in India. Adapted by Buddhists as a solid monument built to enshrine relics of the Buddha, or to mark important sites of Buddhism. In Thailand, the term 'stupa' is sometimes used interchageably with *chedi*.

Sukhothai: The name of an extensive Thai Buddhist kingdom and its capital. The term is also applied to an art style.

Tavatimsa, Dawadeung (T): The heaven of the god Indra, situated at the peak of Mount Meru.

Theravada (P): The School or Teaching of the Elders, a sect of Buddhism considered to adhere most closely to the original teachings of the Buddha. Sometimes called Hinayana, it is the form of Buddhism still practised in Sri Lanka, Burma, and Thailand. *See also Hinayana.*

Tosachat (T): *See Jataka.*

Tosakan (T): *See Ravana.*

Traibhum or *Traiphum* (T): The Three Worlds (divided into 31 realms or levels) of Thai Buddhist cosmology, respectively, from the lowest to the highest: (1) *Kamaphum*: The World of Desire, (2) *Rupaphum*: The World of Form, and (3) *Arupaphum*: The World without Form. Besides being the general term for manuscripts dealing with this subject, it is also the title for mural paintings usually behind the presiding Buddha image in an assembly hall.

Traibhumikatha: The Sermon on the Three Worlds. A Thai Buddhist text.

Ubosot or *Bot* (T): The ordination hall in a Thai monastery, differentiated from an ordinary assembly hall by the presence of *bai sema* markers at the cardinal and sub-cardinal points of the building, indicating consecrated ground.

Vihara (S), *Viharn* (T): An assembly or congregation hall in a monastery compound where religious services, excluding ordination, may be held.

Vishnu (S), *Phra Narai* (T): A major god of Hinduism, worshipped as the Preserver of the Universe in the Hindu Trinity; Vishnu is also responsible for the re-creation of the Universe. In Thailand he is frequently depicted on gable-boards, riding Garuda, or in his incarnation as Rama, hero of the *Ramayana*.

Wat (T): A Thai Buddhist monastery encompassing a variety of religious and secular structures for monks and the religious community.

Yaksha (P), *Yak* (T): A collective term for various classes of celestial demons, usually benevolent, depicted as guardian figures in Thai architectural decoration and painting.

Bibliography

Amnuay Corvanich (1968), *A Brief Note on the Working Elephant*, Bangkok: Forest Industry Organization, Ministry of Agriculture.

_____ (1976), *Thai Elephant*, Bangkok: Forest Industry Organization, Ministry of Agriculture.

'An Account of King Kirti Sri's Embassy to Siam in Saka 1672 [AD 1750]' (1908), translated from the Sinhalese by P. E. Pieris, in *Religious Intercourse between Ceylon and Siam in the Eighteenth Century*, Bangkok; reprinted under the auspices of the Committee of the Vajiranana National Library, by the 'Siam Observer' Office, from *Journal of the Royal Asiatic Society* (Ceylon Branch), 18.

Anuman Rajadhon, Phya (1986), *Popular Buddhism in Siam and Other Essays on Thai Studies*, Bangkok: Thai Inter-Religious Commission for Development and Sathirakoses Nagapradipa Foundation.

Apinpen [pseud.] (1988), 'Chedi Yuthahatthee', *Muang Boran Journal*, 14 (Oct.–Dec.): 68–71, 91.

Beauvoir, Marquis de (1870), *A Week in Siam, January 1867*, A reprint of an extract from the Marquis de Beauvoir's *Voyage autour du Monde* in its original English translation, Bangkok: Siam Society, 1986.

Blofeld, John (1987), *King Maha Mongkut of Siam*, 1972; 2nd edn., Bangkok: Siam Society.

Bock, Carl (1884), *Temples and Elephants: Travels in Siam in 1881–1882*, London: Sampson Low, Marston, Searle, & Rivington; reprinted Singapore: Oxford University Press, 1986.

Boisselier, Jean (1975), *The Heritage of Thai Sculpture*, New York: Weatherhill.

_____ (1976), *Thai Painting*, Tokyo: Kodansha International.

Boonsong Legakul and McNeely, J. A. (1977), *Mammals of Thailand*, Bangkok: Association for the Conservation of Wildlife.

Bowring, Sir John (1857), *The Kingdom and People of Siam*, 2 vols., London; reprinted Kuala Lumpur: Oxford University Press, 1969.

Caddy, Florence (1889), *To Siam and Malaya on the Duke of Sutherland's Yacht* Sans Peur, London: Hurst & Blackett; reprinted Singapore: Oxford University Press, 1992.

Campbell, Reginald (1935), *Teak Wallah: The Adventures of a Young Englishman in Thailand in the 1920s*, London: Hodder & Stoughton; reprinted Singapore: Oxford University Press, 1986.

Caron, François and Schouten, Joost (1671), *A True Description of the Mighty Kingdoms of Japan and Siam*, London; reprinted Bangkok: Siam Society, 1986.

Chadwick, Douglas H. (1991), 'Elephants: Out of Time, Out of Space', *National Geographic*, 179, 5 (May): 2–49.

Cheun Srisavast (1986), 'Raingarnkarnvichaithangwattanatham reuang karnliangchangkhongchaothai-kuay (suay) naichangwatsurin' [Report on Cultural Research Concerning Elephant Care of the Thai-Kuay (Suay) in Surin Province], Cultural Research Project Sponsored by the James H. W. Thompson Foundation, 2528–9 BE (1985–6), Unpublished typescript, held in the Reserve Collection of the Siam Society, Bangkok.

Chirasa Khochachiwa (1986), 'Ganesha Worship and Its Images in Thailand', *Muang Boran Journal*, 12, 2 (Apr.–June): 87–90.

Choisy, Abbé de (1687), *Journal du voyage de Siam fait en 1685 et 1686*, Paris; adapted by M. Dasse, Bangkok: D. K. Book House, 1976.

Chou Ta-kuan (1987), *The Customs of Cambodia*, translated into English by J. Gilman d'Arcy Paul, from Paul Pelliot's French version of Chou's Chinese original, Bangkok: Siam Society.

Chula Chakrabongse, HRH Prince (1960), *Lords of Life: A History of the Kings of Thailand*; 2nd edn., London: Alvin Redman, 1967.

Clark, Kenneth (1977), *Animals and Men*, New York: William Morrow & Co. Inc.

Coedès, George (1968), *The Indianized States of Southeast Asia*, first published in French, 1964; translated by Susan Brown Cowing and edited by Walter F. Vella, Honolulu: East–West Center Press.

Damrong Rajanubhab, HRH Prince (1991), *Journey through Burma in 1936*, translated by Kennon Breazeale, Bangkok: River Books.

Danielou, Alain (1964), *Hindu Polytheism*, New York: Pantheon Books.

Davis, Bonnie (1987), *Postcards of Old Siam*, Singapore: Times Editions.

A Dictionary of Buddhism: Chinese–Sanskrit–English–Thai (1976), Bangkok: Chinese Buddhist Order of Sangha, Wat Pho Maen Khunaram.

Dobias, R. J. (1987), 'Elephants in Thailand: An Overview of Their Status and Conservation', *Tigerpaper*, 14, 1: 19–24.

Doniger O'Flaherty, Wendy (1975), *Hindu Myths: A Source Book Translated from the Sanskrit*, Harmondsworth: Penguin.

Dowson, John (1961), *Classical Dictionary of Hindu Mythology*, London: Kegan Paul.

Fine Arts Department, Thailand (1981), *Khreuang Muhk* [The Art of Mother-of-Pearl Inlay], Catalogue of the Exhibition Commemorating the Centenary Birthday Anniversary of Field Marshal HRH Prince Paribatra of Nagor Svaraga, 29 June 1981, Bangkok.

—— (1988), *Phra Narai Ratchanivet* [The Palace of King Narai at Lopburi], Bangkok: n.p., 1988.

—— (1990), 'Rueangchang' [Concerning Elephants] and 'Reuangchangpheuakkapnangngam' [Concerning White Elephants and Fair Ladies], in *Prachumphrarachaniphondhnairachakarnthi 4 phaapakinaka phaak 1 le Phraphenithambunwankert Phrarachaniphondh Phrabatsomdet Phrachulachomklaochauyuhua* [Miscellaneous Collected Royal Writings of the Fourth Reign, Part 1, and Birthday Meritmaking Commemoration, Royal Writings by His Majesty King Chulachomklao], first published in Bangkok, 1948; reprinted in 1990 from the funeral commemoration of Nai Sampao Pinitka.

—— (1991), *80 Pi Haengkarnanuraksamoradok Thai* [80 Years of Thai Cultural Heritage Preservation], published on the Occasion of the Eightieth Anniversary of the Founding of the Fine Arts Department, Bangkok.

Finlayson, George (1826), *The Mission to Siam and Hue, 1821–1822*, London: John Murray; reprinted Singapore: Oxford University Press, 1988.

Fraser-Lu, Sylvia (1988), *Handwoven Textiles of South-East Asia*, Singapore: Oxford University Press.

Gerini, G. E. (1893), *Chulakantamangala: The Tonsure Ceremony as*

Performed in Siam; rev. edn., Bangkok: Siam Society, 1976.

Gervaise, Nicholas (1688), *The Natural and Political History of the Kingdom of Siam*, Paris; translated and edited by John Villiers, Bangkok: White Lotus, 1989.

Giles, F. H. [Phya Indramontri Srichandrakumara] (1930a), 'Adversaria of Elephant Hunting (Together with an Account of All the Rites, Observances and Acts of Worship to Be Performed in Connection Therewith, as well as Notes on Vocabularies of Spirit Language, Fake or Taboo Language and Elephant Command Words)', *Journal of the Siam Society*, 23, 2: 61–70.

———— (1930b), 'Elephant Hunting in the Korat Table-land', *Journal of the Siam Society*, 23, 2: 71–95.

———— (1932), 'An Account of the Rites and Ceremonies Observed at Elephant Driving Operations in the Seaboard Province of Lang Suan, Southern Siam', *Journal of the Siam Society*, 25: 153–214.

Government Public Relations Office, Thailand (1992), 'Thai Elephant Conservation Centre', in *Thailand Illustrated*, 10, 30 (Sept.–Dec.): 19–24.

Grigson, Henry (1989), *Thai Manuscript Painting*, London: British Library.

Griswold, A. B. and Prasert na Nagara (1971), 'The Inscription of King Rama Gamhen of Sukhodaya (AD 1292), Epigraphic and Historical Studies No. 9', *Journal of the Siam Society*, 59, 2 (July): 179–228.

Haas, Mary (1964), *Thai–English Student's Dictionary*, Stanford: Stanford University Press.

Hallett, Holt S. (1890), *A Thousand Miles on an Elephant in the Shan States*, Edinburgh and London: William Blackwood & Sons; reprinted Bangkok: White Lotus, 1988.

Historical Commission of the Prime Minister's Secretariat (1994), *The Writings of King Mongkut to Sir John Bowring (AD 1855–1868)*, Bangkok. (Advisers: HSH Prince Suphadradis Diskul and MR Supawat Kasemsri; editors: Winai Pongsripian and Theera Nuchpiam; Introduction in Thai by ML Manich Jumsai.)

Hutchinson, E. W. (1940), *Adventurers in Siam in the Seventeenth Century*, London: Royal Asiatic Society; reprinted Bangkok: DD Books, 1985.

Ions, Veronica (1988), *Indian Mythology*, London: Hamlyn, 1967.

Kaempfer, Englebert (1727), *A Description of the Kingdom of Siam, 1690*, London; 1906 edn. reprinted in the Itineraria Asiatica Series, Bangkok: White Orchid Press, 1987.

La Loubère, Simon de (1693), *A New Historical Relation of the Kingdom of Siam*, London; first published in French as *Du Royaume de Siam*, 1691; reprinted as *The Kingdom of Siam*, Singapore: Oxford University Press, 1986.

Larousse Encyclopedia of Archaeology (1984), New York: Hamlyn Publishing Group; first published as *Larousse L'Archeologie decouverte des civilisations disparues*, Paris: Librarie Larousse, 1969; first English edition and translation by Anne Ward, 1972.

Leonowens, Anna Harriette (1870), *The English Governess at the Siamese Court*, London: Trubner; reprinted Singapore: Oxford University Press, 1989.

Livy (n.d.), *The War with Hannibal*, Books 21–30 of *The History of Rome from Its Foundation*, edited with an introduction by Betty Radice, and translated by Aubrey de Selincourt, Harmondsworth: Penguin, 1985.

McDonnell, Etain (1994), 'Jumbo Economics', *Sunday Post* (Bangkok), 20 February.

Manich Jumsai, ML (1991), *King Mongkut of Thailand and the British: The Model of a Great Friendship*, 1972; 3rd edn., Bangkok: Chalermnit.

Marshall, H. N. (1959), *Elephant Kingdom*, London: Travel Book Club.

Mattana Srikrachang et al. (1991), 'Ecology and Numbers of the Asian Elephants in Huai Kha Khaeng Wildlife Sanctuary, Thailand', Unpublished proposal, Royal Thai Forest Department, Bangkok.

Moffatt, Abbot Low (1961), *Mongkut: The King of Siam*, Ithaca: Cornell University Press.

Moss, Cynthia (1987), *Elephant Memories: Thirteen Years in the Life of an Elephant Family*, n.p.: Fontana.

Mouhot, Henri (1864), *Travels in the Central Parts of Indo-China (Siam), Cambodia and Laos during the Years 1858, 1859 and 1860*, 2 vols., London: John Murray; reprinted as *Travels in Siam, Cambodia and Laos, 1858–1860*, Singapore: Oxford University Press, 1989.

Muang Boran Publishing House (1980), *Ayutthaya, The Former Thai Capital*, Bangkok.

_____ (1985), *Mural Paintings of Thailand Series: Khoi Manuscript Paintings of the Ayutthaya Period*, Bangkok.

National Museum Volunteers (1987), *Treasures from the National Museum, Bangkok: An Introduction*, Bangkok.

National Museum Volunteers and Sawaddi Magazine (1979), *The Artistic Heritage of Thailand: A Collection of Essays*, Bangkok.

Neale, F. A. (1852), *Narrative of a Residence in Siam*, London: Office of the National Illustrated Library; reprinted Bangkok: White Lotus, n.d.

Office of the Prime Minister (1982), *Foreign Records of the Bangkok Period up to AD 1932*, Published on the Occasion of the Rattanakosin Bicentennial, Bangkok.

Owens, Patrick and Kulaya Campiranonta (n.d.), *Thai Proverbs*, Bangkok: Darnsutha Press.

Pallegoix, J-B. (1854), *Description du Royaume Thai ou Siam*, Paris; reprinted Farnborough, Hants: Gregg International, 1969.

Payne, Katharine (1989), 'Elephant Talk', *National Geographic*, 176, 2 (August): 261–78.

Piriya Krairiksh (1991a), 'Towards a Revised History of Sukhothai Art: A Reassessment of the Inscription of King Ram Khamhaeng', in James R. Chamberlain (ed.), *The Ram Khamhaeng Controversy: Collected Papers*, Bangkok: Siam Society, pp. 53–159.

_____ (1991b), 'The Date of the Ram Khamhaeng Inscription', in James R. Chamberlain (ed.), *The Ram Khamhaeng Controversy: Collected Papers*, Bangkok: Siam Society, pp. 257–72.

Plion-Bernier, Raymond (1973), *Festivals and Ceremonies of Thailand*, translated from the French by Joann Elizabeth Soulier, Bangkok: Assumption Press.

Plu Luang [pseud.] (1988), 'The Essence of the Treatise on War Strategy', *Muang Boran Journal*, 14, 4 (Oct.–Dec.): 30–6.

Quaritch Wales, H. G. (1931), *Siamese State Ceremonies: Their History and Function*, London: Bernard Quaritch.

_____ (1952), *Ancient South-East Asian Warfare*, London: Bernard Quaritch.

_____ (1977), *The Universe around Them: Cosmology and Cosmic Renewal in Indianized South-East Asia*, London: Arthur Probsthain.

Reynolds, Frank E. and Reynolds, Mani B. (trans.) (1982), *Three Worlds According to King Ruang: A Thai Buddhist Cosmology*, Berkeley Buddhist Studies Series 4, Berkeley: University of California.

Ringis, Rita (1990), *Thai Temples and Temple Murals*, Singapore: Oxford University Press.

Rong Syamananda (1977), *A History of Thailand*, Bangkok: Chulalongkorn University.

Rosenfield, Clare S. (1970), 'The Mythical Animal Statues at Prasat Phrathepphabidon', in Tej Bunnag and Michael Smithies (eds.), *In Memoriam Phya Anuman Rajadhon: Contributions in Memory of the Late President of the Siam Society*, Bangkok: Siam Society, pp. 273–99.

Saeng-Arun Kanokpongchai (1988), 'Treatise on War Strategy and the Thai–Burmese War', *Muang Boran Journal*, 14, 4 (Oct.–Dec.): 26–9.

_____ (1991), 'Indra-Brahma on Rama I Road: Traces of Hindu Beliefs in Thailand', *Muang Boran Journal*, 17, 1 (Jan.–Mar.): 49–56.

Santiapillai, Charles and Jackson, Peter (comps.) (1990), *The Asian Elephant: An Action Plan for Its Conservation*, Gland: IUCN/SSC Asian Elephant Specialist Group.

Sapana Sakya (1994), 'Beast of Burdens', *Bangkok Post*, 28 January.

Segaller, Denis (1982), *More Thai Ways*, Bangkok: Allied Newspapers Limited.

Seni Pramoj, MR and Kukrit Pramoj, MR (1987), *A King of Siam Speaks*, Bangkok: Siam Society.

Sirichai Narumitr (1975), *Old Bridges of Bangkok*, Bangkok: Siam Society.

Sitsayamkan, Luang (1967), *The Greek Favourite of the King of Siam*, Singapore: Donald Moore Press.

Smitthi Siribhadra and Mayurie Veraprasert (1990), *Lintels: A Comparative Study of Khmer Lintels in Thailand and Cambodia*, Bangkok: Siam Commercial Bank and the Department of Fine Arts.

Smitthi Siribhadra and Moore, Elizabeth (1992), *Palaces of the Gods: Khmer Art and Architecture in Thailand*, Bangkok: Asia Books.

Stratton, Carol and McNair Scott, Miriam (1981), *The Art of Sukhothai: Thailand's Golden Age*, Kuala Lumpur: Oxford University Press.

Subhadradis Diskul, MC (1986), *History of the Temple of the Emerald Buddha*, Bangkok: Bureau of the Royal Household.

_____ (1990), *Hindu Gods at Sukhodaya*, Bangkok: White Lotus.

_____ (1991), *Art in Thailand: A Brief History*, 7th edn., Bangkok: Amarin Press.

Subhadradis Diskul, MC and Rice, Charles S. (1982), *The Ramakian (Ramayana) Mural Paintings along the Galleries of the Temple of the Emerald Buddha*, rev. edn., Bangkok: Government Lottery Office.

Sukumar, R. (1989), *The Asian Elephant: Ecology and Management*, Cambridge: Cambridge University Press.

Sullivan, Michael (1984), *The Arts of China*, 3rd edn., Berkeley: University of California Press.

Sunetra Cutinthranon (1988), 'Chakravartin: The Political Concept behind the Thai–Burmese Wars (2091–2397 BE)', *Muang Boran Journal*, 14, 2 (Apr.–June): 89–90.

Suriyavudhi Sukhsvasdi, MR (1988), *Prasat Khao Phnom Rung* (Thai and English), Silpawatthanatham, Art and Culture Series.

Tachard, G. (1685), *A Relation of the Voyage to Siam Performed by Six Jesuits Sent by the French King, 1685*; reprinted Bangkok: White Orchid Press, n.d.

Terwiel, B. J. (1989), *A Window on Thai History*, Bangkok: Editions Duang Kamol.

Thomson, J. (1875), *The Straits of Malacca, Siam and Indo-China: Travels and Adventures of a Nineteenth Century Photographer*, London: Samson Low, Marston, Low, & Searle; reprinted Singapore: Oxford University Press, 1992.

Toke Gale, U (1974), *Burmese Timber Elephant*, Rangoon: Trade Corporation.

Traibhumikatha: The Story of the Three Planes of Existence, by King Lithai (1987), translated by the Thai National Team, led by Khunying Kullasap Gesmankit, Bangkok: ASEAN Committee on Culture and Information.

Van Vliet, Jeremias (1640), *The Short History of the Kings of Siam*, translated by Leonard Andaya from a transcription by Miriam J. Verkuijl-van den Berg and edited by David K. Wyatt, Bangkok: Siam Society, 1975.

_____ (1692), 'A Description of Siam', Leyden (Leiden); reprinted in *Journal of the Siam Society*, 7 (1910).

Vimol Bhongbhibhat et al. (1987), *The Eagle and the Elephant: Thai–American Relations since 1833*, 3rd edn., Bangkok: United States Information Service.

Vincent, Frank, Jr. (1874), *The Land of the White Elephant: Sights and Scenes in South-East Asia, 1871–1872*, New York: Harper & Brothers, Publishers; reprinted Singapore: Oxford University Press, 1988.

Wildlife Fund Thailand: Under the Royal Patronage of HM the Queen (1990), Report and Conservation Activities, Bangkok.

Williams, J. H. (1950), *Elephant Bill*, London: Hart-Davis.

_____ (1953), *Bandoola*, London: Hart-Davis.

Wood, W. A. R. (1924), *A History of Siam*; reprinted Bangkok: Chalermnit Bookshop, n.d.

Wright, Michael (1988), 'The Trading Relationship between Siam and India', *Muang Boran Journal*, 14, 3 (July–Sept.): 88.

Wyatt, David K. (1984), *Thailand: A Short History*, New Haven: Yale University Press.

_____ (1994), *Studies in Thai History*, Chiang Mai: Silkworm Books.

Young, Ernest (1898), *The Kingdom of the Yellow Robe*, London: Archibald Constable & Co.; reprinted Singapore: Oxford University Press, 1982.

Yule, Henry (1858), *A Narrative of the Mission to the Court of Ava in 1855*, London; reprinted Kuala Lumpur: Oxford University Press, 1968.

Zimmer, Heinrich (1946), *Myths and Symbols in Indian Art and Civilization*, Princeton: Princeton University Press.

_____ (1960), *The Art of Indian Asia: Its Mythology and Transformation*, Bollingen Series, XXXIX, edited by Joseph Campbell, New York: Pantheon Books.

Index